Reptiles & Amphibians
of Michigan Field Guide

by Stan Tekiela

Adventure Publications, Inc.
Cambridge, MN

To my wife Katherine and daughter Abigail with all my love

Acknowledgments:

I would like to thank James H. Harding, Instructor/Specialist in Herpetology, Department of Zoology and MSU Museum, Michigan State University, for reviewing the range maps.

Edited by Sandy Livoti

Book and CD design and illustrations by Jonathan Norberg

Range maps produced by Anthony Hertzel

Audio CD recordings by Stan Tekiela and Ritch Tekiela; Fowler's Toad recordings by Jim McGrath

Audio CD produced and edited by Ritch Tekiela of Nature Smart Audio Production Company

Photo credits by photographer and page number:

Cover photo: Female Green Frog by Stan Tekiela
Dudley Edmondson: xii (turtle, scale rows), xiii (both), xiv (frog, toad), 14 (adult head), 24 (both), 26 (head), 38 (belly), 48 (adult, spotted morph), 54 (scale rows), 58 (belly), 86 (terrestrial adult), 88, 98 (both), 108 (brown morph), 112, 118 (brown and red morphs), 122, 124 (green and brown morphs), 128 (female) **James H. Harding**: 40 (adult), 62 (both), 92 **Jeffrey B. LeClere**: 74 (adult, older breeding male), 82, 86 (eft) **Stan Tekiela**: xii (keel, snake, cloaca, keeled and smooth scales, divided anal plate), xiv (foot), 8, 10, 12 (both), 14 (adult, Midland, Midland head), 16 (all), 18 (both), 20 (all), 22, 26 (adult), 34 (both), 36 (both), 38 (adult), 42 (all), 44 (both), 46 (all), 48 (feigning death, head, juvenile), 50 (both), 52 (all), 54 (adult, green morph, red-sided morph), 56 (all), 58 (adult), 60 (both), 64 (all), 66 (all), 68 (all), 74 (juvenile), 76, 84 (all), 86 (aquatic adult), 90 (both), 94, 96, 108 (green morph), 110 (all), 114 (all), 116 (both), 118 (yellow morph), 120, 124 (green brown mix), 126 (both), 128 (male, dark morph), 130, 132 (both) **Robert Wayne Van Devender**: 40 (belly), 100 (both)

Copyright 2004 by Stan Tekiela
Published by Adventure Publications, Inc.
820 Cleveland Street South
Cambridge, MN 55008
1-800-678-7006
All rights reserved
Printed in China

TABLE OF CONTENTS

Introduction

The Reptiles

The Amphibians

MICHIGAN'S REPTILES AND AMPHIBIANS

Reptiles and amphibians are a fascinating, diverse group of animals. Reptiles range from softshell turtles to colorful snakes and darting lizards. Amphibians include silent salamanders and singing frogs and toads. Nearly all are small and harmless, but some reptiles and amphibians still seem to frighten some people. Overall it's a wonderful group that easily holds the attention of young and old alike.

Michigan is a great place for reptiles and amphibians. Every county in the state has at least several species. Chances are you'll be enchanted by these amazing creatures when you spot them!

There is a total of 53 reptile and amphibian species throughout Michigan–30 reptiles and 23 amphibians. Of the 30 reptile species, 10 are turtles, 18 are snakes and 2 are lizards. Of the 23 amphibian species, 10 are salamanders, 11 are frogs and 2 are toads. Unlike many plant and some bird and animal species, all 53 of our reptile and amphibian species are native to the state. Each is as unique and interesting as the next.

HERPETOLOGY

"Herpetology" is the scientific study of reptiles and amphibians. Someone who studies herps is called a herpetologist. "Herp" is an informal term used for any species of reptile or amphibian, and "herpin' " means going out in search of reptiles and amphibians.

Although reptiles and amphibians are not closely related, we don't think twice about grouping them together. Since the number of species in each group is relatively small–over 7,000 reptile and 4,500 amphibian species worldwide compared with more than 9,000 bird and countless plant species worldwide–and because reptiles and amphibians share some common traits, joining the groups seems natural.

All herps have a backbone (vertebra) and are cold-blooded (ectothermic). In ectothermic animals, the surrounding environment regulates body temperature. For example, if it's cold outside the body temperature will be cold. The reverse will be true for the body if the outside temperature is hot. Heating and cooling the body in this way differentiates herps from other animals with backbones (vertebrates) such as mammals and birds, which are warm-blooded (endothermic). Endothermic animals produce heat inside the body and maintain a constant temperature that does not change with the surrounding environment.

Another common trait is dormancy. Reptiles and amphibians both undergo an extended period of hibernation-like sleep in the winter. This inactivity, known as overwintering, can last as long as six months in Michigan, from mid-October to mid-April. Several of the herps also enter a similar dormancy called estivation during hot and particularly dry periods in summer.

Other common traits involve reproduction. The reproductive process of most reptiles and amphibians is accomplished by egg laying and incubation outside of the body. All turtles, lizards and most snakes lay their eggs where sun-warmed soil will incubate them. Frogs and toads lay their masses of eggs in shallow water for warming by the sun.

IS IT A REPTILE OR AN AMPHIBIAN?

"Reptile" is a Latin word that means "to creep or crawl." All of the Michigan reptiles except snakes have feet with well-defined toes and claws. All are covered with scales and have dry skin. While a few snake and lizard species give birth to live young, most of the reptiles lay shelled eggs. Newly hatched young have lungs to breathe air and hatch looking like miniature versions of adults.

"Amphibian" comes from the Greek word *amphibios*, meaning "two lives." Most amphibians have a dual form of existence, breathing one way at the beginning of life and then changing to another. From unshelled eggs immersed in water, most start their lives as water-dependent larvae with gills for extracting air from the water. Later they develop lungs for breathing air directly, transforming to terrestrial life as they develop into adults. Red-backed Salamanders are an exception. Laying their eggs in moist locations (not depositing them into water), their young skip the aquatic stage before hatching into miniature adults. Our largest salamander, the Mudpuppy, is just the opposite. It never leaves the water in any stage of life. Because of its watery habitat, it stays active throughout winter, quite unlike the other amphibians.

Several characteristics set most amphibians apart from most reptiles. Amphibians don't have claws on their toes, they lack scales and lay unshelled eggs. Even though most have internal lungs, their moist skin allows some respiration to occur directly through the skin. In some species, their skin also allows them to secrete a toxic substance to repel predators. Most adult amphibians live near water or in the water.

Amphibians are among the oldest vertebrates to live on the land, evolving approximately 370 million years ago. Frogs evolved about 200 million years ago and haven't changed much since. Salamanders share the long lineage of frogs. They evolved around 170 million years ago—a long time compared with many of the mammals, including humans.

The first reptiles evolved about 40 million years after the first amphibians, around 330 million years ago. The earliest reptiles were small lizards, followed by larger dinosaurs. Snakes evolved approximately 29-38 million years ago. Present-day turtles and lizards appeared about 25 million years ago.

FINDING HERPS

Finding and appreciating herps is an outdoor recreation that is similar to bird watching. Knowing something about the required habitat of an animal will increase your chances of seeing that animal. For example, you wouldn't look for lizards such as secretive Six-lined Racerunners in wetlands or Bullfrogs far from water. With a little bit of effort and some common sense, you can greatly improve your success of finding cool critters such as these.

Turtles

Turtles are found in most ponds, lakes and rivers. They are often observed when they are out of the water, basking in the sun or traveling across land. Turtles are extremely wary while basking and will dive into water when they see the slightest disturbance. Approach these sunbathers carefully, using a pair of binoculars to examine and identify the species from a distance. Turtles walking across land are often females searching for a place to lay eggs. These individuals should not be disturbed, handled or harassed.

Moving any turtle off a busy road, however, is always a good idea. Try to determine which way the animal is going and lend your assistance with minimal disturbance. Help it negotiate street curbs and gutters, which can be major obstacles. Most of our turtles will withdraw into their shells at your approach, but be very careful if attempting to handle a Snapping Turtle. It can harm you even when you are acting in its best interest.

Snakes and Lizards

Most snakes and lizards are usually found in highly specialized habitats. Many live in dry areas with ample hiding places and plenty of food. Carefully lifting or turning over logs, rocks and any other flat debris increases your chances of finding a snake or lizard. Always place the objects back where you found them to maintain the integrity of the original environment.

While snakes and lizards are relatively sturdy critters that can tolerate handling, be respectful if you decide to pick one up. Take care not to inadvertently injure it during handling. Be sure not to drop it if you become frightened. Be especially careful if it is a lizard because if handled adversely it will detach its tail. When you are finished observing or photographing, return the animal to the place where you found it. If you can't identify the species, be even more cautious and observe it without handling.

Michigan has only one venomous snake species. Whenever you are searching a region that has venomous species, such as the

Massasauga, use extra caution when you move objects. **Do not attempt to handle a venomous snake**. If you cannot identify a snake species, it's best to observe it only and not handle it.

Salamanders

Most salamanders live on moist land near ponds, hidden underneath fallen leaves, branches and other forest debris. You can find them by carefully selecting a moist woodland site or wetland and systematically turning over rocks, logs or fallen slabs of bark. Be sure to return each object to its original position to maintain the sheltered habitat it provides.

The best time to see salamanders is after dark on highly humid or rainy evenings. Use a headlamp or flashlight so you can easily see them moving about on the ground. To increase your chances of encountering these critters, look for them during spring and fall migrations (April and September). Be respectful and do not remove any salamanders from the areas where you found them, since they will either be breeding or looking for winter homes.

Salamander skin is much more fragile than the skin of lizards. Salamanders have moist, sensitive skin that dries quickly and absorbs substances on contact. If you've applied insect repellent or sunscreen to your hands, do not handle any of these creatures because the chemicals on your skin can cause them to die.

If your skin is chemical-free, be ready to quickly capture a salamander that may be hiding underneath a rock or a log. Don't worry that it will bite you. Salamanders are nonaggressive and nonvenomous critters.

Some salamanders wiggle madly after capture in an attempt to free themselves. These delicate, often tiny animals can easily be damaged during capture or when in your grasp. Be gentle and keep handling to a minimum. Take care not to accidentally drop the salamander, and be sure to put it back where you found it.

Frogs and Toads

Frogs and toads are easier to find and observe than salamanders. Our frogs are usually found in wet areas such as ponds and small lakes. Toads are seen in a variety of habitats from dry prairies to wetlands. Locating them can be as easy as spotting a Northern Leopard Frog at a local park or an American Toad in your garden. However, since frogs and toads are most active three to four hours immediately after dark, you might have more success spotting them after nightfall using a headlamp or similar flashlight.

The best time to hear most frogs and toads extends from April to July, when the male frogs and toads call to attract mates. These calls are referred to as advertisement calls because the males are "advertising" their readiness to mate. The Frog and Toad Croaking Chart on pages xxii-xxiii shows that all species do not call at the same time. Each species also has a unique call or song, which makes identification possible without actually seeing the individual animal. Since the calls are very important clues to identifying these amphibians, listen to their songs on the accompanying audio CD before scouting around outside. To observe a specific species, listen for the frog or toad song and simply move slowly toward it.

You can easily capture frogs and toads with a small net attached to a length of pole. The guidelines for handling salamanders and the rules of conservation also apply to other amphibians. Refrain from handling frogs and toads when you have insect repellent or sunscreen on your hands, and be sure to return the animal you catch to the area where you found it.

After Handling

Washing up after handling any animal in the wild is a smart, healthy practice. If you have handled a reptile or amphibian, be sure to wash your hands to remove any trace substances that the critter may have excreted. This is a practice you should always do after being out in the field and especially before you eat. There are several brands of waterless washing lotions and towelettes on the market that enable thorough cleansing without the use of water. Using these products is a very practical way to clean up when you are not near running water.

IDENTIFICATION STEP-BY-STEP

Correctly identifying a reptile or amphibian is a relatively easy and straightforward process. Most people have little trouble differentiating a turtle from a snake or a salamander from a frog. Determining what you have begins the identification process.

This field guide is divided into two main sections–reptiles and amphibians. Reptiles are shown first and are divided into groups including turtles, snakes and lizards. Next are the amphibians, shown in two groups of salamanders, and frogs and toads.

Description pages of the groups are illustrated with a silhouette of a turtle, snake, lizard, salamander or frog/toad. Each silhouette is located in a thumb tab in the upper right corner of the page. Once you have decided the type of reptile or amphibian you are seeing, use the thumb tabs to locate the appropriate group. To help speed identification, the individuals within each group are organized by size–from the smallest animal to the largest.

This book provides full-page color photographs to compare with live specimens. Since many herps have unique markings or features, simply examine the photos to make a positive identification.

If you aren't sure of your identification, the text on the description pages explains identifying features that may or may not be easily seen on the photos. Each description page also has a compare section with notes about similar species in this field guide. Other pertinent details and the naturalist facts in Stan's Notes will help you correctly identify your critter in question.

REPTILE AND AMPHIBIAN ANATOMY

It's easier to identify herps and communicate about them if you know the names of the different parts of reptiles and amphibians. For instance, it's more effective to use the term "parotoid glands" to indicate the wart-like bumps behind a toad's eyes than to try to describe them.

Labeled images on the following pages point out the basic parts of reptiles and amphibians.

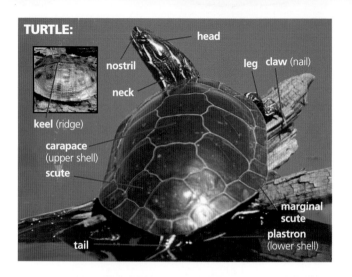

TURTLE:

head
nostril
neck
keel (ridge)
carapace (upper shell)
scute
leg
claw (nail)
marginal scute
plastron (lower shell)
tail

SNAKE:

dorsum (back)
lateral (side)
tail
cloaca (vent)
single anal plate
keeled scales
keel (ridge)
tail
smooth scales
divided anal plate
snout
nostril
head neck
venter (belly)
scales rows
4 3 2 1
belly scales

LIZARD:

- constriction (base) of tail
- lateral (side)
- dorsum (back)
- head
- snout
- nostril
- ear
- leg
- venter (belly)
- tail

SALAMANDER:

- tail
- dorsum (back)
- lateral (side)
- leg
- venter (belly)
- snout
- costal groove
- nostril
- head

FROG:

head
tympanum (ear)
thumb
dorsolateral fold
snout
nostril
upper lip
vocal (throat) sac
lateral (side)
foreleg
toes
front (fore) foot
toe pad
dorsum (back)
hind leg
venter (belly)

TOAD:

head
cranial crests
parotoid glands
dorsum (back)
lateral (side)
nostril
snout
vocal (throat) sac
tympanum (ear)
foreleg
hind leg
toe
venter (belly)

POPULATIONS

The following information about endangered, threatened and special concern species is restated from the Michigan Department of Natural Resources (DNR) main web page, which is located at www.michigan.gov/dnr.

A species is classified as **endangered** if it is near extinction throughout all or a significant portion of its range in Michigan.

A species is **threatened** if it is likely to become classified as endangered within the foreseeable future throughout all or a significant portion of its range in Michigan.

A species is of **special concern** if it is extremely uncommon in Michigan, or if it has a unique or highly specific habitat requirement and deserves careful monitoring of its status. A species on the edge or periphery of its range that is not listed as threatened may be included in this category along with any species that was once threatened or endangered but now has an increasing or protected, stable population.

A species is **stable** if it is not included in the above categories and the population is not declining drastically. A stable species is breeding and reproducing well enough to maintain current populations in a given area.

It is illegal to take, keep, harass, import, transport or sell any part of an endangered or threatened species. Special concern species are not protected by Michigan's Endangered Species Statute or the associated Rules. Please read the full text of the applicable Statute and Rules to understand all regulations pertaining to Michigan reptiles and amphibians that are designated as endangered, threatened or special concern. For further information, please consult the DNR web site.

CAUTION

Whenever you are out in the field, you will need to exercise some level of caution. An encounter with a reptile or amphibian rarely presents danger. Problems usually occur instead from exposure to the weather or contact with toxic plants.

Dehydration, sunburn and/or allergic reaction are a few complications that can result from too much fun in the sun. Protect yourself from ultraviolet rays with sunscreen. Wear a long-sleeved shirt, long pants and a wide-brimmed hat. Be alert to threatening changes in the weather such as lightning or strong winds, and seek safety.

Michigan has more than its fair share of Poison-ivy. Common and widespread, this low-growing plant has groups of three glossy leaves and a woody stem. With more than 90 percent of the population susceptible to the effect of its toxic oil, there are several things you can do to avoid getting a rash. Since it prefers to grow along woodland paths and open fields, stay on paths and avoid needlessly trampling down plants. Reduce your exposure further by wearing boots and clothing that covers your extremities. The best way to lower your vulnerability, however, is to correctly identify Poison-ivy so you can spot it and avoid it before you step in it.

Now a few words about venomous snakes. Venomous snakes are creatures that have evolved powerful toxins for immobilizing prey, not for harming or killing humans. To avoid being bitten by a venomous snake, learn how to correctly identify the species and its habitat, and then stay clear. When in doubt, observe and photograph from a distance. While there are only a few places where rattlesnakes can still be found, do not attempt to pick up any snake that you can't positively identify. Do not try to handle or move any species you don't know. If a snake strikes or bites you, it is acting in self-defense, in fear for its own safety. Seek medical attention immediately, even if you think the bite may be non-toxic. It is always much better to err on the side of caution, since delaying medical attention for a venomous bite can result in death.

Keep in mind that the vast majority of snakes you'll encounter will be nonvenomous. Nonvenomous snakes are quite harmless and will not hurt you, especially if you leave them alone.

As interesting as all of these critters are, resist any temptation to take a reptile or amphibian home. Herps have specific dietary and habitat requirements that are rarely duplicated in a captive situation. Many do not survive captivity, and moving an animal to an unsuitable habitat will most certainly result in its death. The possibilities for future reproduction are also depleted when these animals are removed from the wild. With habitat ranges getting smaller every year, moving just one animal can have a direct impact on the local population of a species. So while herps may captivate you, please don't captivate them. Observe and record them, but do the right thing and leave them where they belong—in the wild!

Enjoy the Herps!

Stan

CALLS OF FROGS AND TOADS

Hearing and identifying frog and toad calls can be a fun and enjoyable learning experience. Identifying frogs and toads by their calls is sometimes easier than seeing a frog or toad. With a little practice you will quickly become familiar with the calls of all 13 frog and toad species of Michigan.

Frogs and toads "advertise" their readiness to mate by making sustained calls in an organized manner known as advertisement calls. Except for very few species, advertisement calls are given by the males. Both males and females of all species, however, are "tuned" to hear their own species calling among many other calling frogs and toads.

Frogs and toads are the only amphibians capable of producing an organized sound. Sound in most species is produced when air forced out of their lungs passes over vocal cords in the larynx. From there it travels into the mouth cavity. Squeezing through tiny holes at the bottom of the mouth, the air enters sacs in the throat region. Throat sacs are the reason why these tiny creatures can produce very loud sounds. The throat sacs inflate with each call and act like resonating chambers, amplifying sound. In some species the sound coincides with the deflation of the throat sacs.

Calling is an intense activity that requires a lot of energy. Studies show that most frog and toad species consume considerably more oxygen when calling than while at rest, and burn more calories than sustained jumping requires. Energy expenditure increases with the rate and duration of calls. After long periods of calling, frogs need to rest or forage for food. Species that call for several successive hours often stop for several hours to rest. Some frogs don't feed at all during the breeding season.

Male frogs and toads call or "sing" to attract females for mating during specific periods in springtime and summer. For example, Western Chorus Frogs start calling in late March and continue to call for more than four months if the conditions are right. The calling season of the Pickerel Frog is much shorter, beginning in early May and ending in mid-June. After the individual breeding season ends, the species falls silent until the following spring.

Depending upon the species, most male frogs and toads call from the water's surface or edge alone or in groups of up to several thousand. However, if you hear a call that doesn't sound exactly like any of the recordings on the audio CD, it doesn't mean it is not one of our 13 species or that you have discovered an entirely new species. Environmental factors, such as time of season or air temperature, and individual variations, such as size or location, influences the volume, pitch, speed, cadence or even the number and frequency of the calls. Aggression or territorial calls often sound completely different, and this just adds to the confusion.

The beginning of the calling season of some frogs and toads, such as Wood Frogs and American Toads, works like a switch flipping on. In other species, they start calling one by one. Toward the end of their calling season most also stop one by one; however, some individuals will continue to call.

Changes in air temperature can affect the number of calls and how often you'll hear them. Fewer males call on cold nights and more call on warmer nights. While each species is different, many frog and toad species won't call when the air temperature is below 40°F (4°C) on early spring nights or 50°F (10°C) during summer. The few frogs and toads that call on nights with even lower air temperatures usually have calls that don't sound like their normal calls, such as being slower or less frequent.

Size, age and vigor of the individual also makes a big difference in the quality of the call. Just as people have different singing voices, individual frogs and toads have variations in their calls. For example, the male with the best perch often sounds louder and clearer than the surrounding males. The call of this male may have more resonance and carry farther than the calls of other frogs. The dominant male gives the initial call after a quiet period and starts the chorus calling.

The volume and intensity of frog and toad calls varies among the species. Some have very soft calls that attract individuals nearby. Others give loud calls to attract females as far as a mile away. It has been suggested that species living in permanent ponds and lakes have quieter calls to attract mates living in the same body of water. Species such as the Western Chorus Frog, which call from ephemeral breeding ponds in different locations each year, have a loud call for long-distance communication.

Another factor that influences the volume is the perch or location of a frog or toad. Some frogs call while floating on the surface of the water. Others call while clinging to vegetation several inches above the water's surface. Treefrogs frequently call while they are in trees. The sound of a treefrog calling high up in a tree carries farther than the call of the same frog sitting at the edge of water in thick vegetation. Pickerel Frogs often call with their heads above water with their resonating vocal sacs at the water's surface, which reduces the volume. Interestingly, Pickerel Frogs will sometimes call while completely submerged underwater, greatly reducing the volume. Another factor affecting volume is the direction in which the frog or toad is facing.

Geographic variations or "dialects" also cause variations in frog and toad calls. Like songs of some bird species, the calls of some frog and toad species aren't identical in different parts of the state. Dialects reflect genetic differences among breeding populations. While these variations may be subtle, they can be heard with a discriminating ear.

The Frog and Toad Croaking Chart on the following pages will help you determine which of our Michigan frogs and toads you are hearing. The species are listed in chronological order based on when the males start to call. The entire calling season, which extends from late March to mid-August under normal conditions, is organized into early, mid- and late seasons. These categories are only general guidelines, however. Variations in nature make it possible to hear some frogs and toads calling out of season or in other unusual ways. Both males and females can also produce distress or warning calls, and release calls when grasped by a predator or another frog. You can hear all of these calls on the accompanying audio CD.

FROG AND TOAD CROAKING CHART

13 Michigan Species	Calling Season	
	MARCH	**APRIL**
EARLY SEASON (2-3 days after ice is off breeding ponds)		
Western Chorus Frog (pg. 109)		
Spring Peeper (pg. 113)		
Wood Frog (pg. 119)		
Northern Leopard Frog (pg. 125)		
MID-SEASON (tree leaves emerging, flowers blooming)		
Gray Treefrog (pg. 117)		
Cope's Gray Treefrog (pg. 115)		
American Toad (pg. 133)		
Fowler's Toad (pg. 131)		
Pickerel Frog (pg. 121)		
LATE SEASON (warm nights, plenty of insects)		
Cricket Frog (pg. 111)		
Green Frog (pg. 127)		
Mink Frog (pg. 123)		
Bullfrog (pg. 129)		

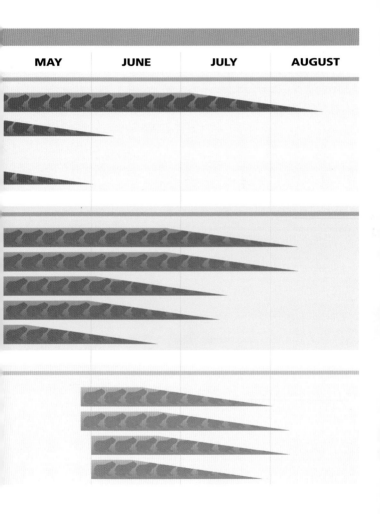

MAY	JUNE	JULY	AUGUST

Common Name
Scientific name

Section
(GROUP)

Family: common family name (scientific family name)

Size: average range of upper shell length (turtles), average range of length from snout to tip of tail (snakes, lizards, salamanders), average range of length from snout to rump (frogs, toads), may include (M) male and (F) female sizes

Description: brief description of the reptile or the amphibian, may include color morphs or young

Eggs/Young: number of eggs laid or offspring produced and when; hatchling, live birth, larva or tadpole; when hatchlings emerge, offspring are born or larvae or tadpoles transform; additional comments

Origin/Age: native or non-native to Michigan; average life span in the wild

Habitat: environment where the animal is found (e.g., fields, prairies, deciduous forests, coniferous forests, ponds, lakes, wetlands); specific locations, when applicable

Overwinter: where and how the animal spends the winter

Food: what the animal eats most of the time (e.g., insects, earthworms, mammals, fish, frogs, spiders, slugs)

Compare: notes about other species that look similar, and the pages on which they can be found

Stan's Notes: Interesting gee-whiz natural history information. This can be something to look or listen for, or something to help positively identify the animal. Also includes remarkable features.

Description pages have one of these designations to indicate the status of the population. See page xv for descriptions of these classifications.

STATUS: Endangered, Threatened, Special Concern or Stable

Turtles

THREATENED
Spotted Turtle

SPECIAL CONCERN
Eastern Box Turtle
Wood Turtle
Blanding's Turtle

STABLE
Common Musk Turtle
Common Map Turtle
Painted Turtle
Spiny Softshell Turtle
Red-eared Slider
Snapping Turtle

Turtles are long-lived reptiles that have flourished for more than 200 million years. Most species have an average life span of over 25 years and some live longer than half a century. Hardy animals, they are unique because they have a shell. While most species are hard-shelled, others have a soft shell.

In hard-shelled turtles, the upper shell (carapace) is formed from bones fused to the ribs and vertebrae. The lower shell (plastron) is formed from bones fused to the ribs and shoulders. These shells are connected at each side by bone and cartilage.

Overlying the bony structure are a series of semiround to square, domed scales called scutes. When a hard-shelled turtle ages, a new layer of shell that is slightly larger than the old forms beneath the scutes. Hardening of the new shell forces old scutes to peel off like large flakes of skin. This often goes unnoticed because the shell of a turtle is usually wet, which hides the peeling process just like moisturizer helps hide flaking skin on hands.

In some hard-shelled species you can estimate the age by counting scute growth rings. When a scute peels off, it leaves a ring around the edges where it attached to the shell. Just like the rings of a tree, the number of scute rings increases with time. Counting these rings is much easier to do when a turtle is young. It is nearly impossible to determine the age of an old turtle because the rings on the shell are often obscured by years of wear and tear.

In softshell turtles, much of the bony part of the shell is replaced with a leather-like skin. Soft shells grow in the same way as hard shells, but they're pliable and leathery to the touch and lack scutes.

The shell of a turtle is its main defense against most predators. It is well known that a turtle can withdraw its legs and head into its shell. It usually wraps its tail up around the outer surface since it is unable to withdraw it. Some turtles, such as the Blanding's Turtle, have a hinge on the front quarter of the lower shell. Once the animal withdraws inside, the hinge allows the lower shell to close very tightly and seal off the opening to its most vulnerable spot–its head.

It is not possible for a turtle to leave its shell. While the shell is an outstanding defense against predators, the protection provided by it is no match for an automobile. Nearly all turtles hit by cars will not survive. The force usually fractures the shell, crushes internal organs and results in death.

The common name "Turtle" is often applied to all aquatic species. "Tortoise" is often reserved for the land-dwelling animals. There are no tortoise species in Michigan. There are many differences between turtles and tortoises. Turtles have flat, webbed feet for swimming. Tortoises have thick, stubby feet that are better suited for moving around on land and digging into earth. Aquatic turtles, especially the males, often have long sharp claws. Males use their claws to hold onto females when mating. In some species, length of the claws on the front feet is one way to differentiate between the sexes.

Aquatic turtles are air-breathing animals that need to surface for oxygen. At the surface, they will poke only their eyes and nostrils above water to look around and take a breath before they head back down. Some aquatic turtles have a special lining in the mouth and throat that extracts oxygen from water and augments respiration, working like a modified gill. Remaining submerged for extended periods of time, most will eat and sleep underwater and many overwinter underwater for months. Some bury themselves in silt and mud at the bottom, while others will wedge themselves beneath underwater logs and rocks. Unlike some frog and toad species, the bodies of adult turtles don't freeze during winter. Ice fisherman routinely report seeing turtles swimming under the ice in winter. However, much more needs to be discovered about the winter activities of our turtles.

Most turtles are slow when traveling on land, but they can make short bursts of quick movement to escape from predators or catch prey. Their smooth shells and webbed feet are much better suited for quickly slipping through water and swimming very fast. Turtles hunt by eyesight and smell. Fish eaters, such as Snapping Turtles, chase after small prey or lie in wait, quickly extending their necks to grab an unsuspecting fish swimming close by. Softshell turtles are capable of swimming faster than some species of fish. They rely on speed to help capture their fish prey.

Turtles consume a variety of other foods, from frogs and worms to plants and insects. All species of turtles are toothless but have sharp-edged jaws called a beak. Like birds, turtles use their beaks to grab and tear prey or cut vegetation. Many feed on crayfish, snails and dead fish or animals, using their beaks to cut chunks of meat free from carcasses or to crack shells.

All turtle species have excellent eyesight and, for the most part, they are silent. Some may give a short hiss.

Turtles are cold-blooded (ectothermic) animals. Most of the species spend a good part of the day warming themselves, basking in the sun. The sunlight helps promote greater muscle activity, increase metabolism and fight infection and disease. Basking also helps remove algae and parasites that have attached to the turtle while underwater. Basking is especially important for the Painted Turtle and other species that eat a lot of plant matter. It speeds digestion and helps the production of vitamin D.

The instinct to bask is very strong. Jockeying over choice basking spots usually results in large numbers of turtles piling on top of each other. With an optimal body temperature of about 78-82°F (26-28°C), turtles orient themselves to receive the maximum amount of solar radiation for quickly raising their temperature. Once that temperature is reached, they often return to feeding or resting in the water.

Courting and mating occur in spring. The males of many turtle species are well adapted for mating. They have a concave area on their lower shell (plastron) that fits over the domed upper shell (carapace) of the female. Males of some species also have extra long claws on their front feet to stroke the sides of a female's head during courtship. This behavior helps entice the female to breed.

After mating, females of most turtle species leave the water to dig "nests" in soft, often sandy soils for depositing eggs. A female will choose sandy banks along rivers and lakes or loose soils in gardens and lawns, where the sun-warmed earth unobstructed by trees will incubate the eggs. Using her hind legs and excavating backward, she tunnels 6-12 inches (15-30 cm) down and digs a chamber large enough to deposit her eggs. She covers the hole with dirt before leaving and makes little attempt to conceal the site. All of this is done without seeing what she is doing since she digs facing away from the nest.

Most species of turtles lay leathery, round white eggs that bounce and don't crack when dropped into the egg chamber, and harden later. Most turtle species in Michigan have eggs about the size of a Ping-Pong ball or smaller.

Once a female has covered her eggs and left the nest site, she never returns to see her young or offer any protection. Turtle eggs are preyed upon by many mammals and bird species such as gulls. It is estimated that more than 70 percent of the nests are preyed upon by mammals such as skunks, foxes, coyotes, bears and raccoons. In some regions this figure can reach closer to 90 percent. Mammals look for recently disturbed soils and use their sense of smell to find the nests. After the mammals have had their fill, the gulls usually clean up. Even after hatching, the baby turtles are preyed upon by other animals. The small percentage of hatchlings that do survive to adulthood usually live a long time and reproduce often.

In many species of turtles, the nest temperature during incubation determines the sex of offspring. Generally speaking, warmer nests incubated at around 84-86°F (29-30°C) produce mostly females. Cooler turtle nests at 76-77°F (24-25°C) produce mostly males. Nests incubated at 82-84°F (28-29°C) produce an even mix.

Similar to baby birds, a hatchling turtle has a pointed egg tooth (caruncle) for breaking open the eggshell. Most hatchlings emerge from their earthen nests in August or September and immediately head for the nearest water. Offspring of some species hatching late in September or October often remain underground until the following spring before digging out and heading for water. It is possible that these hatchlings partially freeze during winter since the nests are too shallow to escape the frost of winter.

Most turtles seen on the road in summer are females searching for places to dig nests and deposit eggs. Some journey several miles to find suitable soils to dig their egg chamber. However, most travel only a couple hundred feet to find a suitable site. Turtles on the road in autumn are males and females moving to places to spend the winter.

Common Musk Turtle
Sternotherus odoratus

Turtles
(REPTILE)

Family: Musk and Mud Turtles (Kinosternidae)

Size: 3-5" (7.5-13 cm)

Description: Narrow, domed olive-to-gray upper shell (carapace). Small lower shell (plastron) with an inconspicuous hinge. Large head with a pointed snout. Two light stripes on each side of head. A blunt, horny nail on the end of male's tail. Female has a short tail.

Eggs/Young: 1-10 eggs once per year; hatchling; August and September; sex is dependent upon soil temperature

Origin/Age: native; 50 or more years

Habitat: permanent water bodies such as ponds, lakes, rivers

Overwinter: underwater; burrows into mud beneath submerged logs, overhanging riverbanks and muskrat houses

Food: aquatic insects, snails, crayfish, small fish, tadpoles, aquatic plants

Compare: Usually is smaller than the more common Painted Turtle (pg. 15), which has a flatter upper shell and many prominent yellow or red stripes on its neck. Blanding's Turtle (pg. 25) also has a high-domed carapace, but it has a distinctive yellow chin.

Stan's Notes: Rarely away from the water. Basks on logs and rocks in shallow water. In hot weather it moves to deeper water. Overwinters in shallow water when water temperature drops. Secretes a musky substance from glands on sides, discouraging predators. Sometimes called Stinkpot. Preyed upon by raccoons, skunks and otters. Male matures in 3-5 years, female in 8-10 years. Mates in early April and May. Female lays eggs in a nest cavity or sometimes under logs, often barely covered. Up to 80 percent of nests are predated. Incubation lasts 60-90 days. Hatchlings grow to full size quickly. Head and neck stripes often fade on older adults.

STATUS: Stable

Spotted Turtle
Clemmys guttata

Turtles
(REPTILE)

Family: Pond and Box Turtles (Emydidae)

Size: 3-5" (7.5-13 cm)

Description: Many small yellow and orange spots on upper shell and skin. Cream-to-yellow lower shell, edged with black blotches. Male has brown eyes, a tan-to-dark chin and a long thick tail. Female has orange-to-red eyes, yellow chin and a short, thinner tail than male.

Eggs/Young: 3-8 eggs once per year; hatchling; August and September or overwinters underground until the next spring; sex is dependent upon soil temperature

Origin/Age: native; 30 or more years

Habitat: permanent water bodies such as ponds, small lakes, wet or boggy areas, slow rivers with mud bottoms

Overwinter: burrows into mud in shallow water

Food: worms, crayfish, aquatic insects, small fish, tadpoles, aquatic plants such as algae, water lily, duckweed

Compare: Blanding's Turtle (pg. 25) is larger and lacks spots. Look for a small, dark turtle with irregular yellow and orange spots. Some Spotted Turtles have spots on the head and neck, but lack spots on the shell.

Stan's Notes: Most often seen in early spring, basking in sun. Rarely seen in summer. Moves when shallow breeding ponds dry up, and becomes dormant (estivates) when it gets too hot. Usually in shallow ponds with Painted Turtles. Also walks on land. Male matures at 7 years, female at 10-14 years. Males often fight during mating season, establishing dominance. Mates in March and April; some mate in fall. Young nearly double in size their first year. Populations have declined over the past 50 years due to habitat loss (draining of ponds and wetlands) and reptile collectors, who value the unique shell.

STATUS: Threatened   **11**

Common Map Turtle
Graptemys geographica

Turtles
(REPTILE)

Family: Pond and Box Turtles (Emydidae)

Size: M 4-6" (10-15 cm); F 7-12.5" (18-32 cm)

Description: Carapace is olive green to brown with a network of thin yellow lines (resembling contour lines on a map). Indistinct ridge (keel). Saw-toothed marginal scutes near tail. Lower shell is pale yellow to creamy. Skin is brown to olive with thin yellow (sometimes green-to-orange) lines similar to, but more pronounced than the lines covering the carapace. A single small yellow dot or spot behind each eye. Female frequently has a smooth carapace with dark blotches.

Eggs/Young: 10-20 eggs up to twice a year; hatchling; September or next spring; sex is dependent on soil temperature

Origin/Age: native; presumed 20-25 years

Habitat: large rivers

Overwinter: underwater in groups with other map turtle species; wedges in logjams or tree roots under exposed banks

Food: clams, crayfish, snails, aquatic insects, fish, plants

Compare: Painted Turtle (pg. 15) has bold yellow stripes on head and neck and red edges on its shell. Look for a small yellow spot behind each eye to help identify the Common Map Turtle.

Stan's Notes: A highly aquatic turtle that prefers large rivers with quiet backwaters and oxbows. Its specialized diet of clams and snails makes this turtle unique. Male has more pronounced shell marks and midline keel than female, a thicker, longer tail and longer front claws. The larger female usually has faint yellow lines on its upper shell and is well known for its large powerful jaws, which can crack open a clam or snail. Male matures at 3-5 years, female at 10-14 years.

STATUS: Stable 13

Midland

Painted Turtle
Chrysemys picta

Turtles
(REPTILE)

Family: Pond and Box Turtles (Emydidae)

Size: M 4-7" (10-18 cm); F 4-9" (10-22.5 cm)

Description: Upper shell is overall dark olive green to nearly black with red edges. Yellow lower shell, tinged with red. Head, neck, legs and tail are olive green to nearly black with bold yellow stripes. Midland has bold red stripes on neck and forelegs.

Eggs/Young: up to 20 eggs once or twice a year, starting in June; hatchling; 70-80 days or the following spring; sex is dependent upon soil temperature

Origin/Age: native; 15-25 or more years

Habitat: lakes, ponds, rivers, creeks, permanent water sources

Overwinter: underwater in lakes and large ponds; rests on bottom or burrows into mud

Food: aquatic insects, fish, crayfish, snails, tadpoles, plants

Compare: The Blanding's Turtle (pg. 25) has a yellow chin and a plastral hinge. Look for red markings on plastron to identify the Painted/Midland Turtle.

Stan's Notes: Our most common turtle. Very cold tolerant, it emerges early in April, often when ice is still on lakes. Can be seen swimming under ice in late winter. Basks in sun in large groups. Slips into water if approached. Female matures at 6-10 years. Male uses front claws to stroke and hold female while courting and mating. Female travels to south-facing slopes or loose soil to lay eggs. Not uncommon for young to hatch, remain in the nest and dig out the following spring. Place a section of chicken wire or hardware cloth over a nest site in your yard and anchor it to the ground. This will help prevent skunks, raccoons or dogs from digging up the eggs. Use 2-inch (5 cm) wire so hatchlings can climb through to exit.

STATUS: Stable 15

lower shell

male

female

Eastern Box Turtle
Terrapene carolina

Turtles
(REPTILE)

Family: Pond and Box Turtles (Emydidae)

Size: 4-8" (10-20 cm)

Description: High-domed upper shell, variable color and pattern, but usually dark brown to black (sometimes olive) with dull yellow or orange spots, blotches and lines. Some appear all yellow with dark marks. Low keel. Dark lower shell with yellow or orange blotches is hinged at front third, allowing for tight closure. A large head with yellow or orange. Male has orange-to-red eyes. Female has brown or dull yellow eyes.

Eggs/Young: 3-12 eggs once per year; hatchling; 50-90 days or next spring; sex is dependent upon soil temperature

Origin/Age: native; 50 or more years

Habitat: wet meadows, pastures, fields

Overwinter: burrows in soil and leaf litter down to 2 feet (.6 m), some individuals burrow only a few inches, others dig less, exposing the carapace

Food: plants, earthworms, slugs, fruit, mushrooms

Compare: Blanding's Turtle (pg. 25) has a yellow chin and lacks yellow blotches and spots on upper shell.

Stan's Notes: A long-lived terrestrial turtle, rare to uncommon in the state. Diurnal, moving in early morning and late afternoon. Spends hot middays partially buried in leaf litter. Small home range of 4-40 acres (1.6-16 ha). Males wander more than females, some finding their way home from a half mile away. Most are killed by cars, skunks and raccoons. Consumes a wide variety of fruit such as strawberries and raspberries. A major seed dispersal agent. Was collected and eaten by early settlers and Native Americans. Converting woodlands to farmlands eliminated box turtles from much of its former range.

STATUS: Special Concern **17**

Spiny Softshell Turtle
Apalone spinifera

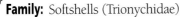

Turtles
(REPTILE)

Family: Softshells (Trionychidae)

Size: M 5-9" (13-22.5 cm); F 7-18" (18-45 cm)

Description: Smooth, leather-like, flexible, oval upper shell with many tiny spines on front edge near neck. Overall light brown to olive green with many dark circles on male and blotches on female. White-to-gray lower shell (plastron). Head and legs match carapace color. Yellow chin and line through eyes. Dark markings on neck and legs. Long tubular snout with 2 projections in the nostrils. Webbed toes.

Eggs/Young: up to 40 eggs once or twice a year; hatchling; August and September

Origin/Age: native; 30 or more years

Habitat: large rivers, lakes and ponds, preferably with sand or mud bottoms

Overwinter: underwater in river bottoms

Food: aquatic insects, small fish, frogs, crayfish, snails

Compare: The only smooth, leather-like turtle in the state. Look for the long tubular snout to help identify.

Stan's Notes: One of the most recognizable turtles in Michigan. One of a few softshell species in North America. River turtle, but found in larger lakes with shallow areas and sandbars. Can stay underwater for long periods, pumping water over linings in throat, where it extracts oxygen from water. Swims faster than many fish, which helps it catch some. Very shy. Retreats to water if disturbed. Adult female is larger than male and has blotches on upper shell. Male has a longer, thicker tail. Male matures at 5 years, female at 8-10 years. Mates in May. Lays eggs near water or on a sandbar. Young male and female look similar with dark circles on carapaces. Dormant October to April.

STATUS: Stable

Wood Turtle
Glyptemys insculpta

Family: Pond and Box Turtles (Emydidae)

Size: 5-9" (13-22.5 cm)

Description: Highly sculpted upper shell, tan to dark brown with lighter radiating lines. Yellow lower shell (plastron) with dark blotches. Upper surface of exposed skin is brown with areas of yellow or orange. Lower surface is all yellow or orange.

Eggs/Young: up to 18 eggs once per year; hatchling; August and September depending upon soil temperature

Origin/Age: native; 30-40 or more years

Habitat: rivers and streams with woodland floodplains

Overwinter: underwater beneath ice, often in riverbanks

Food: plants, berries, mushrooms, earthworms, slugs

Compare: The Blanding's Turtle (pg. 25) has a hinged plastron. Look for Wood Turtle's highly sculpted carapace and yellow or orange areas on skin.

Stan's Notes: An uncommon species in Michigan. While most turtles need to feed underwater, relying on water pressure to help swallow, this species is terrestrial, feeding mostly on land. This is a diurnal (daytime) turtle that was once collected for food and biology experiments and as pets. Species name *insculpta* means "sculptured" and refers to its shell, which appears as though it were sculpted from a block of wood (hence its common name). Female matures at 15-20 years, but is most reproductive when in her 30s and 40s. Emerges in April to mate. Lay eggs in riverbanks and sandbars. Like a few other turtle species, sex of hatchlings apparently is not determined by soil temperature. Counting rings on scutes to estimate age usually works only when a Wood Turtle is under 25 years. Shells become smoother in older turtles.

Red-eared Slider
Trachemys scripta

Turtles
(REPTILE)

Family: Pond and Box Turtles (Emydidae)

Size: 5-11" (13-28 cm)

Description: A brown-to-olive upper shell with yellow and black lines and spots. Olive green head, neck, legs and feet with bold yellow lines. Characteristic red dash mark behind eyes. Yellow lower shell with black blotches. Old males can be very dark (melanistic) with dark pigment covering the yellow and red marks.

Eggs/Young: 3-30 eggs once per year; hatchling; 65-80 days or overwinters underground until the following spring; sex is dependent upon soil temperature

Origin/Age: native, may have been introduced; 30 or more years

Habitat: permanent bodies of water with aquatic vegetation, lakes, reservoirs, swamps, backwaters of large rivers

Overwinter: burrows into mud at the bottom of lakes or beneath submerged logs or muskrat lodges

Food: aquatic insects, snails, crayfish, fish, tadpoles, algae, duckweed and other aquatic plants, some seeds

Compare: Painted Turtle (pg. 15) has red markings on edges of its upper shell and lacks red marks behind its eyes.

Stan's Notes: Named for the red mark behind the eyes where ears are expected to be. Red mark may be less defined or missing in some. Called "Slider" for its habit of sliding from basking sites into water. Feeds early in the morning and late in the day, basking on large logs and rocks in water during the warmest hours, stacking on each other when sites are full. Male is smaller than female, has greatly elongated front claws and a longer tail. Hatchlings are 1-1.5 inches (2.5-4 cm) long with bright green skin and yellow stripes. Was unjustly persecuted due to the misbelief that it kills and eats many game fish.

STATUS: Stable   23

Blanding's Turtle
Emydoidea blandingii

Turtles
(REPTILE)

Family: Pond and Box Turtles (Emydidae)

Size: M 6-11" (15-28 cm); F 5-9" (13-22.5 cm)

Description: A very dark, smooth, high-domed carapace. Hinged lower shell (plastron) is yellow with dark blotches along the outer edges. Upper surface of exposed skin is dark brown to black. Lower surface is bright yellow. Large head and long neck. Yellow throat, chin and lower jaw. A notched upper jaw makes it appear as though it is smiling.

Eggs/Young: up to 20 eggs once per year, June; hatchling; August and September

Origin/Age: native; 50 or more years

Habitat: ponds, small lakes, wetlands, quiet rivers

Overwinter: underwater in muddy lake and pond bottoms

Food: crayfish, snails, insects, frogs, worms, plants, fish

Compare: The Painted Turtle (pg. 15) has a network of yellow or red markings on neck. Eastern Box Turtle (pg. 17) also has a high-domed shell and hinged plastron, but its upper shell has distinct markings. Look for Blanding's distinctive yellow throat and chin.

Stan's Notes: A gentle turtle that rarely bites. Male is larger than the female, with a longer tail and concave lower shell (plastron). The hinged plastron allows it to withdraw its head and legs and close up tightly. Able to feed on land and underwater, unlike most of the other Michigan turtles. Cold tolerant. Can be seen swimming underneath the ice in late winter. Emerges in April to mate. Female travels up to a mile to find a sandy open location to lay eggs. Older females in their 40s and 50s have more reproductive success. Hatchlings sometimes must travel great distances to find water.

STATUS: Special Concern **25**

Snapping Turtle
Chelydra serpentina

Turtles
(REPTILE)

Family: Snapping Turtles (Chelydridae)

Size: M 12-19" (30-48 cm); F 10-15" (25-38 cm)

Description: Dark green-to-brown (sometimes black) upper shell (carapace) with large, pointed, toothed rear marginal scutes. Younger turtles have 3 prominent serrated keels. Lower shell is uniquely small and light yellow to white. Large head, thick neck and powerful jaws. Webbed feet and long sharp claws.

Eggs/Young: up to 30 eggs once a year; hatchling; 60-130 days or the next spring; sex is dependent on soil temperature

Origin/Age: native; 30 or more years

Habitat: lakes, ponds, rivers, creeks, permanent water sources

Overwinter: underwater; burrows into mud

Food: aquatic insects, carrion, fish, frogs, crayfish, snails, salamanders, plants

Compare: Look for the large, thick head and neck and large, pointed, toothed rear marginal scutes of Snapping Turtle. Prominent ridges on upper shell of younger individuals. Long tail and a greatly reduced lower shell (plastron).

Stan's Notes: Largest turtle in Michigan. Some individuals can weigh 50 pounds (23 kg) or more. Crepuscular with few natural predators. Defends itself by turning and facing to strike and bite. Its long neck allows it to reach a distance that is half the length of upper shell. An omnivore, eating plants and animals. Highly aquatic species, feeding while completely submerged and allowing water pressure to aid in swallowing. Male is larger than female and has a longer tail. Female matures at about 12 years. May travel a great distance to find a good area to lay eggs.

STATUS: Stable **27**

Snakes

ENDANGERED
Kirtland's Snake
Copper-bellied Water Snake

THREATENED
Eastern Fox Snake

SPECIAL CONCERN
Massasauga
Black Rat Snake

STABLE
Red-bellied Snake
Brown Snake
Ring-necked Snake
Butler's Garter Snake
Smooth Green Snake
Queen Snake
Eastern Hognose Snake
Northern Ribbon Snake
Common Garter Snake
Milk Snake
Northern Water Snake
Western Fox Snake
Racer

There are more than 3,000 species of snakes in the world today. Our modern-day snakes, which have been around for about 5-10 million years, evolved from lizards as far back as 150 million years ago. The majority of snakes are found in warm climates near the equator. The farther north you are from the equator, the fewer species you will see. In Michigan we have only 18 snake species.

Some snakes in Michigan are small and thin. Most are only 3 feet (.9 m) or less in length and under 1 inch (2.5 cm) in width. Many of our most common species, such as the Red-bellied, Brown and Ring-necked Snakes, measure less than 15 inches (38 cm) and are as thin as a pencil. Our largest snake, the Black Rat Snake, can reach a length of nearly 6 feet (1.8 m) and width of approximately 2 inches (5 cm).

The skin of snakes is not slimy or wet, but soft and dry to the touch, and is periodically shed as a snake grows. The actual shedding takes only a few minutes, but the time leading up to a shed can take as long as two weeks. The first indication a snake is about to shed is a slight change in color. The eyes will appear as though covered with a bluish film and the skin will look lackluster. A snake in this condition is in the "blue" stage. The blues can last a couple days to a week. Just before shedding, the bluish in the eyes will disappear and the snake will return to a relatively normal appearance. The actual shedding starts when a small area of skin loosens, usually near a corner of the mouth. The snake will rub the loosened patch of skin against an object such as a stick or rock. When the skin catches on the object, the snake will use the resistance to move away from it and slither out, leaving the old skin behind. The old skin rolls back, peeling off the body inside out–in the same way a sock turns inside out when peeled off your foot.

Snakes are unique reptiles because they have a body shape that allows them to adapt to different habitats, from arid deserts to tropical rain forests. The backbone (vertebra) of most snakes consists of 150-400 interconnecting bones, enabling extreme flexibility. Each Michigan species has one row of extra-wide belly scales that extends from side to side along the underside of the body from head to tail. Each belly scale is attached to internal muscles, which, when contracted and expanded, propels the snake forward.

The rest of a snake's body is covered with rows of smaller, uniform scales. Each body scale is attached at one end and slightly overlaps the next, allowing the skin to contract and expand without bunching. There are two types of body scales–keeled and smooth. Keeled scales have a small ridge or keel along the length of each scale, while smooth scales do not. Body scales can sometimes help identify some snake species or even a shed skin.

If you run your finger along the length of a snake from the head toward the tail, it will feel smooth. Go the opposite way and you will feel individual scales catch your fingertip. The rear edge of the belly scales is what grips a surface and provides enough friction for movement. The edge of a scale is also capable of gripping smooth surfaces. All scales working together give snakes the remarkable ability not only to move with agility, but to climb vertical objects such as trees.

Snakes usually move by contracting their belly muscles, but they also contract muscles that cause them to move from side to side, in an S pattern. This is commonly referred to as slithering. They may also push against rocks and sticks to help them maneuver along on the ground. Side-winding is another method of moving. This type of movement has the least contact with the ground and is practical when the ground is too hot to be comfortable. Most snakes employ several methods of locomotion.

Snakes can maneuver rather quickly and will flee from danger rather than stay and fight. The fastest snakes are only able to move about 2-4 miles (3-6 km) per hour—slower than a human can walk, so you can easily outrun a snake. We do not have snakes in North America that can outrun humans or any that will attempt to attack or chase you. All of our snakes are shy and just want to be left alone.

Snakes are quite good swimmers, and use the same muscle contractions for slithering to move across the water's surface. While they usually hold their heads several inches above the surface, they will occasionally dive below. Some will rest on the bottom to escape danger. Others can maneuver faster in water than on land.

All species of snakes have a discernible head, neck, body and tail. In many, the head is the same width as the neck. Some have a wider head than neck. Either way, the neck is the flexible, short region behind the head. The body extends from just behind the neck to the vent. The vent, located on the underside, is an external opening for waste elimination and reproduction. It appears as a slit between belly scales. Next to the vent is a belly scale known as the anal plate. There are two types of anal plates—divided and single (undivided). The type of anal plate can often confirm the identity of a snake species or a shed skin.

Many people have a hard time telling the difference between the body and tail. Snakes may appear to be "all tail," but the tail is just the region from the vent to the pointed end or tip. Tail length, which is measured from vent to the tip, is often the best way to differentiate male from female. Males generally have longer tails.

All snakes have teeth, but do not use them for chewing. Some species have larger teeth than others. Venomous species have two specialized, very long teeth called fangs, designed to deliver a small amount of powerful venom. Most snakes have several rows of small sharp teeth in the upper jaw. These teeth have a slight backward curve, which helps a snake hold wiggling prey and swallow. Many species can swallow prey up to two times the size of their own heads. This is possible because the lower jawbone can unhinge (disarticulate), allowing the mouth to open wide. The bottom jaw also expands at the front to permit the passage of a large meal. Most snake species take only a few minutes to grab, kill and swallow prey of average size.

Snake eyes are covered with a clear scale. The eye scale (spectacle) provides good protection and also keeps each eye moist. Lacking eyelids, snakes cannot blink and must sleep with their eyes open.

Rattlesnakes and other venomous snakes have sensory openings known as heat pits (facial pits) between the eyes and nostrils. We refer to this variety of snakes as pit vipers because of their heat pits. Heat pits give these snakes the ability to detect heat (infrared radiation), which helps them find prey in light and dark conditions.

Even though all snake species have nostrils, their sense of smell is experienced with the tongue. A snake's flicking, darting tongue indicates that it is sampling the air and transferring trace chemicals from the surroundings into its Jacobson's organ–seven small organs in the roof of its mouth. From there the information is transferred to the brain. The tongue in snakes is forked, which allows them to distinguish between left and right and provides more surface area for sampling air.

Snakes have a reduced ability to hear because they lack external openings for ears. Instead, they sense vibration and low frequency sounds through their jawbones and belly scales. While humans can detect sound from great distances, a snake can feel vibrations only a few feet away.

For the most part, snakes are silent. Some snake species can make a hissing noise by forcing air out of their lungs, which produces a rather loud hiss. This noise is generally enough to warn anyone disturbing them to stay away. Rattlesnakes have a series of rattles on the tips of their tails that they shake and vibrate, creating a characteristic sound.

All snakes are carnivores that feed mostly on live animals. They generally swallow their animal prey headfirst and whole, without chewing. Snakes frequently find prey by following a scent, which often leads into a burrow or tunnel. They usually kill their prey by suffocation either by constriction, a process of wrapping a body coil around their victim and squeezing, or by swallowing the prey live. Some snake species will lay and wait for something edible to pass within reach, and quickly extend themselves just enough to seize the meal with their mouths. Small snakes will actively seek slugs and worms. Venomous snakes will bite once before releasing their prey, then wait for the venom to take effect before feeding.

Many snake species, such as the Milk Snake, reproduce by laying eggs (oviparous). The female will excavate a small depression under a log or rock, or burrow into loose or sandy soils to deposit her eggs. Depending upon the species, she will lay 5-30 small, round white eggs ranging from the size of a jellybean to a small chicken egg. After laying the eggs, she relies on the sun and the warmth of the earth to incubate them. Other snakes, such as the Common Garter Snake, give birth to live young (ovoviviparous).

Snakes are generally diurnal (daytime) animals. They can also be active at night during hot weather in summer. Sometimes they take advantage of warm days in late autumn to bask in sunlight before heading underground for the winter. Some snakes emerge early in springtime if the weather is warm enough, but they retreat underground again when the cold returns.

Snakes use well-established burrows or dens below the frost line. Even though snakes don't have a social structure and are solitary by nature, they often den together with other snake species. It is possible to see many snakes together in early spring when they emerge from these dens.

The occurrence of venomous snakes in Michigan is so low that your chances of seeing any of these animals is rare. Since many snake species vibrate their tails when they're excited or alert, many people mistake a harmless, nonvenomous snake for a rattlesnake. It is important to note that all snakes are an integral part of any ecosystem despite what some people may think. Snakes are a good predator to have around because they're good at controlling populations of mice and other small mammals, which can easily overpopulate and cause problems.

In general, snakes will not harm you and will do nearly anything to stay away from humans. Snakes will bite only when backed into a corner or if they are roughly handled. The teeth of a snake are extremely small and fragile, the jaws are not very powerful and rarely will a bite to a human result in much more than a break in the skin. If you should happen to sustain a snakebite, however, stay on the safe side and seek medical attention right away.

belly

Red-bellied Snake
Storeria occipitomaculata

Family: Colubrid Snakes (Colubridae)

Size: 8-10" (20-25 cm)

Description: A thin, small snake with a bright red or orange belly. Overall brown, gray or nearly black. Most have 2-4 dark thin stripes on back (dorsum) and along each side (lateral). Some have a lighter center stripe. Top of head is often dark brown. Head is about as wide as body with a thinner neck. White chin and throat. Keeled scales, divided anal plate.

Eggs/Young: up to 15 offspring once per year; live birth; August and September; young 2-4" (5-10 cm) long; light in color, light mark on neck, lighter belly than adult; matures at 2 years

Origin/Age: native; presumed 4-8 years

Habitat: deciduous forests, fields

Overwinter: moist deciduous forests that are often close to water, yards, fields, prairies

Food: earthworms, slugs, snails, insect larvae

Compare: Brown Snake (pg. 37) has a cream-to-pink belly and is lighter brown than some Red-bellied adults. Often seen in association with Brown Snake and Common Garter Snake (pg. 55). Look for the bright red or orange belly to help identify.

Stan's Notes: A harmless, pencil-sized woodland snake, too small to give a serious bite. Small enough to be eaten by many animals. Good for yards due to its appetite for slugs. Like Brown Snakes, has special teeth to extract snail bodies from shells. Male has a longer tail than female. Active days and summer nights. Often killed on roads during migration to and from winter dens. Emerges in late April.

STATUS: Stable   35

belly

Brown Snake
Storeria dekayi

Snakes
(REPTILE)

Family: Colubrid Snakes (Colubridae)

Size: 9-12" (22.5-30 cm)

Description: Overall light brown to gray with 2 parallel rows of dark spots running down the back (dorsum). Spots may connect, forming a ladder-like pattern. Belly (venter) is cream to light pink. Head is only slightly wider than body. Small dark markings under each eye. Larger dark spots on back of head. Dark stripe behind each eye. Keeled scales, divided anal plate.

Eggs/Young: up to 40 offspring once a year; live birth; 3-4 months, mostly July and August; young 3-4" (7.5-10 cm) long

Origin/Age: native; presumed 5-10 years

Habitat: moist deciduous forests that are often close to water, yards, fields, prairies

Overwinter: underground below the frost line

Food: earthworms, snails, slugs, small frogs

Compare: Often confused with the Red-bellied Snake (pg. 35), which has a much redder belly and lacks dark spots on back of head.

Stan's Notes: A very secretive snake with a small home range of less than a couple hundred yards. Often spends much of its time underground or below leaf litter. Feeds heavily on earthworms and snails, using its specialized teeth for removing snails from shells. Finds prey by its sense of smell. Small thin snake that is eaten by many other animals because of its size. Male is slightly smaller than female, but has a longer tail. Emerges in late April to feed and mate. Male finds female by following a pheromone odor she releases. Young grow quickly, maturing at 2 years. Returns to same winter den each year.

STATUS: Stable   37

belly

Ring-necked Snake
Diadophis punctatus

Snakes
(REPTILE)

Family: Colubrid Snakes (Colubridae)

Size: 10-15" (25-38 cm)

Description: Shiny gray to dark blue or black. An obvious yellow or orange ring around neck. Yellow belly changes to orange, then red nearer the tail with few dark marks on belly. Smooth scales, divided anal plate.

Eggs/Young: up to 10 eggs once a year, June and July; hatchling; 40-60 days; young 3-4" (7.5-10 cm) long; hatches looking like adult

Origin/Age: native; presumed 4-8 years

Habitat: bluffs and rock outcroppings with an abundance of vegetation (often facing south), moist woodlands

Overwinter: underground mammal dens below the frost line

Food: earthworms, slugs, insects, amphibians

Compare: Brown Snake (pg. 37) and Red-bellied Snake (pg. 35) lack a yellow or orange ring around neck. Look for a yellow belly changing to orange, then red near tail.

Stan's Notes: A small, secretive snake, preferring hillsides with rocks and thick vegetation. Rarely seen due to its nocturnal habits. Also known as the Corkscrew Snake, it will coil, hide its head and expose its yellow to orange to red underside. It emits a foul-smelling odor when handled, but rarely bites. Hunts mostly by sense of smell. Will sometimes constrict to kill prey. Female matures at 2-3 years. Mates in May. Lays eggs in moist soils under rocks and rotten logs. Several females may lay eggs together in an area that was used many times before. Overwinters communally with other snakes. This species is common on some of the large islands in Lake Michigan. Also called Northern Ring-necked Snake.

belly

Kirtland's Snake
Clonophis kirtlandii

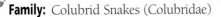

Family: Colubrid Snakes (Colubridae)

Size: 10-24" (25-60 cm)

Description: Overall reddish brown (sometimes gray) with dark, often indistinct blotches in 4 rows down the back and sides (lateral). Pink-to-red belly with a row of small black spots along each side. A small, narrow, dark head, often black. White, cream or yellow lips and chin. Keeled scales, divided anal plate.

Eggs/Young: 5-15 offspring once per year; live birth; August and September

Origin/Age: native; 5-10 or more years

Habitat: wet prairies and fields, fens, swales, often near water

Overwinter: underground dens and tunnels, crayfish burrows

Food: earthworms, slugs, leeches

Compare: The Red-bellied Snake (pg. 35) is much smaller and lacks dark spots on back. The Ring-necked Snake (pg. 39) is also smaller, with a ring around its neck.

Stan's Notes: This is a secretive snake, rarely seen. A high association with water, but not an aquatic species like Queen Snake. Believed to spend much of its time below ground in tunnels or under leaf litter. May wiggle violently and try to hide its head under a loop of its body before dashing for cover when disturbed. Often flattens its body and becomes rigid when threatened. Male is smaller than female, but has longer a tail. Babies grow quickly and mature at 2 years. Reaches its northernmost limits in southern Michigan. Appears to be restricted to a relatively small region bounded by Michigan, Illinois, Kentucky and Ohio. Highly dependent upon wet grassy meadows. Population has dropped dramatically due to agricultural practices and housing project construction. A very difficult snake to keep in captivity.

STATUS: Endangered 41

belly

scale rows 1 2 3 4 5

belly scales

Butler's Garter Snake
Thamnophis butleri

Snakes
(REPTILE)

Family:	Colubrid Snakes (Colubridae)
Size:	15-20" (38-50 cm)
Description:	Overall brown to black or olive. Lateral stripes are light yellow to orange yellow. May have a double row of black spots between stripes. Small dark head. Thick neck. Keeled scales, single anal plate.
Eggs/Young:	10-25 offspring once per year; live birth; August and September
Origin/Age:	native; 5-10 or more years
Habitat:	old fields, roadsides, vacant lots, marshes, prairies
Overwinter:	underground mammal dens and tunnels, crevices, building foundations
Food:	earthworms
Compare:	Common Garter (pg. 55) has side stripes on second and third scale rows, unlike Butler's lateral stripes on third scale row and adjacent halves on second and fourth scale rows (see inset). Butler's has a smaller head and thicker neck than Common Garter Snake.

Stan's Notes: Very easy to confuse with the Common Garter Snake. Will attempt to bite when first handled, but otherwise not aggressive. Like other garters, will release an unpleasant musky secretion when frightened. Locally common in the southeastern part of the Lower Peninsula. Wide variety of habitats but rarely in woodlands, like the Common Garter. Active from April through mid-October, sometimes November. Mates right after emerging from winter dormancy. Only about one-half to three-quarters of females mate each year. Females give birth in mid- to late summer, with larger females giving birth to more young. Matures in 2 years.

belly

Smooth Green Snake
Opheodrys vernalis

Snakes
(REPTILE)

Family: Colubrid Snakes (Colubridae)

Size: 12-24" (30-60 cm)

Description: A smooth-scaled, overall bright green snake with a white or pale yellow belly (venter). Head is slightly wider than body. White lips and chin. Black-tipped red tongue. Smooth scales, divided anal plate.

Eggs/Young: up to 10 eggs once per year; hatchling; 10-30 days; young 4-6" (10-15 cm) long and olive green

Origin/Age: native; presumed 4-10 years

Habitat: fields, prairies, along forest edges, open woodlands

Overwinter: underground dens below the frost line

Food: grasshoppers, crickets and other insects, caterpillars, centipedes, spiders

Compare: The only bright green snake in Michigan. The Racer (pg. 67) has a bright yellow belly.

Stan's Notes: A remarkably colored snake. Also called Grass Snake because of its color and grassy habitat. A harmless snake that is too small to inflict a painful bite. Escapes by blending in or turning and fleeing when threatened, but will coil and strike if cornered. The only Michigan snake that feeds nearly exclusively on insects, hunting for them by eyesight during the day. Male has a longer tail than female. Matures at 2 years. Emerges in April. Mates in May. Female lays eggs with other females in a communal nest in rotting vegetation or decaying logs, where decomposition heats and incubates the eggs. Female can retain eggs in the body for an extended period, shortening the incubation time needed after depositing eggs in the nest. Overwinters with other snake species underground. Populations have declined due to habitat loss and use of insecticides, which kill its food source. Bright green color quickly fades to dark blue after death.

STATUS: Stable

belly

side

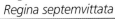

Queen Snake
Regina septemvittata

Snakes
(REPTILE)

Family: Colubrid Snakes (Colubridae)

Size: 12-30" (30-76 cm)

Description: Overall tan to brown (sometimes olive) with a light yellow-to-white lateral stripe on the second scale row and upper half of the first scale row. Pale yellow belly with 4 brown stripes, 2 on the belly, 2 on sides, on the first half of the first scale row. Lateral stripes often fade in older adults. A small head. Yellow-to-white chin and throat. Keeled scales, divided anal plate.

Eggs/Young: 5-30 offspring once per year; live birth; July through September

Origin/Age: native; presumably 5-10 or more years

Habitat: clear, shallow, rock-bottomed streams and rivers with an abundance of crayfish, also ponds, lakes, ditches

Overwinter: underground in mammal dens or crayfish tunnels, usually near water

Food: crayfish, small fish, tadpoles

Compare: Redbelly Snake (pg. 35) lacks the light lateral stripes and has a red or orange belly. Ring-necked Snake (pg. 39) lacks any lateral stripes and has an obvious yellow or orange ring around its neck.

Stan's Notes: A slender water snake and an excellent swimmer. Active day and night, sometimes moving up into branches or bridges over water during the day, dropping down into the water when disturbed. Feeds nearly exclusively on crayfish that have molted their hard shells. Probes for soft-bodied crayfish beneath rocks and other submerged objects, using its tongue to detect ("smell") prey. Does not do well in captivity because of the specialized food. Also susceptible to declining populations due to its unique habitat and food requirements.

STATUS: Stable

feigning death

spotted morph

juvenile

Eastern Hognose Snake
Heterodon platirhinos

Snakes
(REPTILE)

Family: Colubrid Snakes (Colubridae)

Size: 15-30" (38-76 cm)

Description: Variable in color and markings (morphs). Uniformly gray to brown with 2 large dark spots just behind the head and a series of faded dark spots on the center of back (dorsum). Belly (venter) is pale yellow to gray with underside of tail lighter than belly. Snout turns up like a shovel. Gray-to-brown adult spotted morph has well-defined spots or blotches. Keeled scales, divided anal plate.

Eggs/Young: up to 30 eggs once per year; hatchling; August and September; hatchling may have a dark belly with a lighter chin and tail, bold marks on juvenile can fade

Origin/Age: native; presumed 20 or more years

Habitat: woods, floodplains, fields or prairies with sandy soil

Overwinter: underground dens or will burrow below frost line

Food: mainly toads, also salamanders and frogs

Compare: The Hognose Snake's turned-up snout and defensive posture–raised head, flattened neck and hissing, then feigning death–make it easy to identify.

Stan's Notes: A thick-bodied digging snake. Uses its unique snout to dig in loose soils. Pushes its nose in earth and sweeps back and forth while moving forward. A diurnal hunter, finding prey by sight and odor, then seizing it with enlarged rear teeth. Also called Puff Adder, Hissing Adder, Sand Adder or Spreadhead. Raises head and flattens neck like a cobra, hissing and striking with mouth closed if cornered. Feigns death if threatened, sometimes vomits and turns over, belly up with mouth open and tongue exposed. If rolled onto its belly, it rolls back and plays dead again. "Comes back to life" when safe.

STATUS: Stable 49

belly

Northern Ribbon Snake
Thamnophis sauritus

Snakes
(REPTILE)

Family: Colubrid Snakes (Colubridae)

Size: 15-30" (38-76 cm)

Description: Long and slender snake, dark brown to black. Lateral stripes are light yellow to white, on third and fourth scale rows. Brown stripe on the first and second scale rows. A long tail, about one-quarter to one-third total body length. Head is long, narrow and wider than the neck. White lips. Keeled scales, single anal plate.

Eggs/Young: 5-25 offspring once per year; live birth; July and August

Origin/Age: native; 5-10 or more years

Habitat: sphagnum bogs, wetlands, along ponds

Overwinter: underground mammal dens and tunnels, crevices

Food: fish, frogs, salamanders, tadpoles

Compare: Common Garter Snake (pg. 55) has a stouter body and shorter tail. Common Garter has yellow lateral stripes on second and third scale rows up from belly scales, while the Northern Ribbon has lateral stripes on the third and fourth scale rows.

Stan's Notes: A garter snake found throughout the Lower Peninsula. Populations extend from Michigan to Maine and south to Florida. A semi-aquatic species, almost always in wet areas. Basks in sun and escapes to water when startled. Not very aggressive, but will try to bite when first handled. The female is only slightly larger and thicker bodied than the male. Will release an unpleasant musky secretion if frightened, like other garter snakes.

heat pit

segmented rattle

Massasauga
Sistrurus catenatus

Snakes
(REPTILE)

Family: Viperid Snakes (Viperidae)

Size: 15-48" (38-76 cm)

Description: Thick-bodied rattler, light gray to brown with 20-40 dark spots or blotches on back. Smaller alternating spots on sides, often outlined with a thin white line. Tail appears banded with dark markings. Dark stripe from each eye to neck. Two stripes on top of head to neck. Vertical pupils. Heat (facial) pits. A very small, segmented dark rattle. Keeled scales, single anal plate.

Eggs/Young: 3-20 or more offspring once every 2-4 years; live birth; August and September; young are often 7-11" (18-28 cm) long

Origin/Age: native; presumed 20-25 or more years

Habitat: prairie wetlands, swamps, woodlands, river valleys

Overwinter: dens, crayfish burrows or crevices below frost line

Food: small mammals, birds, snakes, amphibians

Compare: Resembles Western Fox Snake (pg. 65) and Eastern Hognose Snake (pg. 49). Look for a small rattle on a banded tail. Note range and rarity.

Stan's Notes: This is the only venomous snake species in the state. The elliptical, catlike pupils of venomous snakes differentiate them from nonvenomous snakes, which have round pupils. Has been hunted and killed because it is venomous. Hunts in daytime, but gets more active at night during hot spells in summer. Finds prey by sensing body heat through its heat (facial) pits and vibration. Common name "Massasauga," a Chippewa Indian word meaning "great river mouth," presumably indicates the habitat in which it was found. Overwinters by itself or with other Massasaugas. Exceedingly rare in Michigan and not found in the Upper Peninsula.

STATUS: Special Concern 53

green morph

scale rows 1 2 3 4

belly scales

red-sided morph

Common Garter Snake
Thamnophis sirtalis

Snakes
(REPTILE)

Family: Colubrid Snakes (Colubridae)

Size: 16-40" (40-102 cm)

Description: Highly variable in color (morphs). Dark and slender snake with 3 yellow (sometimes green, orange or tan) stripes along body. Middle stripe on back (dorsum) usually lighter than side (lateral) stripes. Some have prominent red dashes or hash marks between dorsal and lateral stripes. Pale yellow belly. Dark head. Often a yellow or pale green upper lip. Nearly white chin. Keeled scales, single anal plate.

Eggs/Young: 10-25 offspring once per year; live birth; August and September

Origin/Age: native; 5-10 years

Habitat: just about any habitat from prairies to wetlands

Overwinter: underground mammal dens, crevices, foundations

Food: amphibians, worms, minnows, slugs, small animals

Compare: Common Garter's side stripes are on the second and third scale rows (see inset) compared with Butler's (pg. 43) lateral stripes on the third scale row and adjacent halves on the second and fourth scale rows.

Stan's Notes: The most common snake in Michigan. Variable in color with different colored stripes and backgrounds. "Garter" refers to its yellow lateral stripes, which resemble stripes on garter belts. Follows scent of prey to find food. Will bite when handled, usually resulting in a minor scratch. Emits a strong musky scent as a defense. Often seen in spring in large, swirling mating masses of up to 20 snakes consisting of males and one female (often the biggest). Female starts to reproduce in her second year. Overwinters in large groups, sometimes with different snake species. Mistakenly called Garden Snake. Also known as Eastern Garter Snake.

STATUS: Stable

belly

Milk Snake
Lampropeltis triangulum

Snakes
(REPTILE)

Family: Colubrid Snakes (Colubridae)

Size: 24-42" (60-107 cm)

Description: Variable in color and markings. Most are uniformly gray or tan with a row of large brown-to-red saddle-shaped markings on center of back, each outlined in black. A row of same color smaller spots along each side. Belly is white or cream with a checkerboard of dark spots. Relatively small head, white chin and a V- or Y-shaped dark mark on back of neck. Smooth scales, single anal plate.

Eggs/Young: up to 20 eggs once per year; hatchling; 30-40 days; young up to 10" (25 cm) long; looks like adult with brighter red marks

Origin/Age: native; 10-15 years

Habitat: deciduous woods, valleys, ridges, rocky areas

Overwinter: underground below the frost line

Food: mice, voles, lizards, small snakes, birds, bird eggs

Compare: This is the only Michigan species with brown, red or orange saddle-shaped markings outlined in black. Western Fox Snake (pg. 65) lacks the black outlines around its markings and has a reddish brown head.

Stan's Notes: A slender constrictor, often under rocks and logs on hills in spring and fall, moving to woods and farms in summer. Attracted to rodents in barns, but named after the notion that it drinks milk from cows. When disturbed, coils and strikes while vibrating its tail. Often killed because its bright spots remind some of poisonous species. Secretive, remaining hidden most of the time. Overwinters in groups, emerging in mid- to late April. Mates in May. Lays eggs in decaying woodpiles or underneath flat rocks. Smaller snakes are an important prey item for young Milk Snakes.

STATUS: Stable

belly

Northern Water Snake
Nerodia sipedon

Snakes
(REPTILE)

Family: Colubrid Snakes (Colubridae)

Size: 24-42" (60-107 cm)

Description: Overall reddish brown to tan or gray. Appears to be all brown, especially when dry. Dark brown-to-black bands on neck and tail. Large, often saddle-shaped marks on the back (dorsum) with alternating lateral spots. Belly is cream-colored and covered with many reddish half-moon marks with dark outlines. Keeled scales, divided anal plate.

Eggs/Young: up to 40 offspring once per year; live birth; August and September; young 10" (25 cm) long; appears similar to adult; matures at 2-3 years

Origin/Age: native; presumed 10-15 years

Habitat: permanent water sources, ponds, lakes, rivers

Overwinter: underground mammal burrows below the frost line, crayfish tunnels or rock crevices

Food: fish, frogs, salamanders, tadpoles, crayfish

Compare: Often confused with venomous snake species. Seen in similar habitat as Massasauga (pg. 53), but lacks the small dark rattle. Look for Water Snake's reddish half-moon markings on belly to help identify.

Stan's Notes: Rarely away from water. Frequents structures near water such as boathouses, piers, bridges and outhouses. Often basks in sun on logs and rocks. Releases a foul smell if threatened, discouraging any predators. Escapes to lake bottom when disturbed. If cornered on land, defends itself by flattening its neck and head and striking. Able to give a painful, nonpoisonous bite. Mistaken for the deadly Water Moccasin (Cottonmouth) and killed, but Cottonmouth doesn't occur in Michigan. Doesn't affect game fish populations. Emerges in April.

STATUS: Stable

belly

Copper-bellied Water Snake
Nerodia erythrogaster

Snakes
(REPTILE)

Family: Colubrid Snakes (Colubridae)

Size: 24-55" (60-140 cm)

Description: Overall dark brown to black. Head is usually darker than the body. Bright orange lips and chin. Belly is orange to red (rarely yellow) with small dark spots along each side. Keeled scales, divided anal plate.

Eggs/Young: 5-30 offspring once per year; live birth; September and October

Origin/Age: native; 5-15 or more years

Habitat: ponds, lakes, slow-moving rivers and streams, flood-land forests, oxbows

Overwinter: dry underground dens and tunnels, crayfish burrows

Food: prefers frogs and salamanders, also eats crayfish, fish

Compare: The Northern Water Snake (pg. 59) has dark spots on its back, lacks the orange-to-red belly and is more associated with water.

Stan's Notes: A large snake, often basking in sun on a log near water. Quickly retreats to water if disturbed and can stay submerged for up to an hour. When cornered or captured, it often flattens its neck and discharges feces and smelly musk. Known to repeatedly strike and bite, but is nonvenomous. Diurnal except for hot parts of summer, when it becomes nocturnal or inactive (estivates) under logs or rocks until it is cool. Male is often smaller than female and has a longer tail. Juveniles have paler bellies than adults. They also have a pattern of dark blotches, which quickly fade as they grow. This species is rarely seen away from water except during migration, when it moves to dry upland sites to overwinter. Also moves in summer if its pond or stream dries. Declining throughout its range, especially in Michigan. Endangered due to habitat destruction and changing climate.

STATUS: Endangered   **61**

belly

Eastern Fox Snake
Elaphe gloydi

Snakes
(REPTILE)

Family: Colubrid Snakes (Colubridae)

Size: 24-60" (60-152 cm)

Description: Overall tan to yellow (sometimes light brown) with many (average 34) large, dark brown-to-black spots on back and smaller spots down sides. Belly is tan to yellow with dark blotches. Head is often reddish or light orange. Weakly keeled scales, divided anal plate.

Eggs/Young: 5-30 eggs once per year in June and July; hatchling; 60-78 days depending upon the soil temperature; young 6-10" (15-25 cm) long

Origin/Age: native; presumed 20 or more years

Habitat: shorelines of ponds, rivers, marshes, dunes, beaches

Overwinter: underground dens and tunnels below the frost line

Food: small mammals, birds, bird eggs

Compare: Western Fox Snake (pg. 65) is nearly identical, but it has more dorsal blotches (average 41). Use range to identify since they are separated geographically.

tan's Notes: Often associated with water and sandy soils. A diurnal unter that finds prey by eyesight and movement. Feeds mostly on mall mammals, quickly grabbing with its mouth and squeezing with everal coils until prey is suffocated. Swallows prey much larger than s head. Vibrates tail when alarmed. Often mistaken for a rattlesnake. Occasionally confused with Eastern Copperhead, a venomous snake hat is not in Michigan, because of its reddish orange head. Not easy o tell male from female. The male may be slightly larger and have a proportionally longer tail than female. When female is carrying eggs gravid), she often has a noticeably thicker body and more abruptly apering tail. Of the two fox snake species in Michigan, the Western ox Snake is more widespread and abundant than the Eastern.

STATUS: Threatened 63

juvenile

belly

Western Fox Snake
Elaphe vulpina

Snakes
(REPTILE)

Family: Colubrid Snakes (Colubridae)

Size: 24-60" (60-152 cm)

Description: Overall brown to tan with many (average 41) large dark brown spots or saddle-shaped blotches on the middle of back (dorsum) and alternating with smaller spots on sides. Belly is pale yellow with many brown spots. Usually unmarked head, often light brown to reddish brown to copper, and widest just behind the eyes. Round snout. Keeled scales, divided anal plate.

Eggs/Young: up to 30 eggs once a year, June and July; hatchling; 60-78 days depending upon the soil temperature; young 6-10" (15-25 cm) long

Origin/Age: native; presumed 20 or more years

Habitat: fields, prairies, woodlands and floodplains that are not far from water, river bottoms

Overwinter: underground dens below the frost line

Food: small mammals, birds, bird eggs

Compare: Northern Water Snake (pg. 59) is darker with many half-moon reddish marks on the belly.

tan's Notes: Excellent climber, but usually on the ground. Diurnal unter. Seizes and holds prey, squeezing with several coils until it's lead, then swallows it whole. Confused with rattlesnakes, vibrating ts tail when alarmed. Mistaken for the deadly Copperhead (doesn't occur in Michigan) and needlessly killed. May release a musky, fox-ike odor when captured, hence its common name. Male is slightly arger than female and has a longer tail. Emerges from communal lens in April. Lays eggs under logs in loose soil. Juvenile is gray with brown dorsal spots outlined in black, with dark marks on top of ead. Also called Spotted Adder. Known as Pine Snake in Michigan.

juvenile

belly

Racer
Coluber constrictor

Snakes
(REPTILE)

Family: Colubrid Snakes (Colubridae)

Size: 30-60" (76-152 cm)

Description: Long, slender snake with smooth, shiny scales. Slate gray with bright blue sides and belly, but varies in color from black and gray to blue and yellow. Some have a white belly. Narrow dark head, same width as the neck. White chin and large eyes. Smooth scales, divided anal plate.

Eggs/Young: up to 25 eggs once a year, June and July; hatchling; 50-65 days depending on soil temperature, August and September

Origin/Age: native; 5-10 years

Habitat: open dry prairies and fields, bluffs, outcroppings

Overwinter: underground mammal dens below the frost line

Food: mammals, snakes, lizards, frogs, insects, birds, eggs

Compare: The Smooth Green Snake (pg. 45) is bright green. Black Rat Snake (pg. 69) shares the white chin, but has marks on belly and lacks Racer's slate blue color.

Stan's Notes: Also known as Blue Racer. Named for being quick and agile and for its blue color. Not a constrictor as the species name implies. Holds its head 6-8 inches (15-20 cm) above ground to see well as it moves. Uses eyesight to locate prey. Quickly gives chase, seizes prey with mouth and "chews" it to death. Has a large range, up to 20 acres (8 ha). Heads for shelter if threatened. Coils, strikes and vibrates its tail if cornered. Releases a foul substance if handled or injured, so not a good pet. Male has a longer tail than the female. Active during the day (diurnal) starting in April. Mates in May and June. Hatchlings and juveniles have black-bordered spots on gray or brown bodies. Overwinters with many snakes in underground dens.

black morph

juvenile

belly

yellow morph

Black Rat Snake
Elaphe obsoleta

Family: Colubrid Snakes (Colubridae)

Size: 36-70" (90-178 cm)

Description: A large, nearly all-black snake with white, yellow or red flecks. White chin and throat. Gray belly (venter) with dark spots or a checkerboard pattern. Body is squared in cross section. Head is distinctive from the body, often darker and wider just behind the eyes. Keeled scales, divided anal plate.

Eggs/Young: up to 30 eggs once a year, June and July; hatchling; 60-80 days; young 10" (25 cm) long

Origin/Age: native; presumed 20 or more years

Habitat: woodlands, forested hills, rock outcroppings

Overwinter: underground burrows below the frost line

Food: mice and other small mammals, birds, bird eggs

Compare: Much darker than the Racer (pg. 67), which has a distinctive blue belly.

Stan's Notes: This is the largest snake in Michigan. Not very common in the state and on the decline. Also known as Pilot Blacksnake due to the erroneous belief that it led rattlesnakes and other snakes to den sites for winter. Adept at climbing trees, which is how it obtains birds and bird eggs. This is a diurnal (daytime) snake that uses its eyesight and sense of smell to hunt. Harmless to people. Does a great job controlling small mammal populations. Will coil, vibrate tail and strike when it is approached. Emerges in April and May. Matures at 3-5 years. Lays eggs in hollow logs or loose soil. Gray with large dark spots, young do not resemble adults. They gradually change color, appearing dark at 2-4 years.

Lizards

There are over 3,300 lizard species found around the world. Although most are concentrated in tropical and desert regions, we have only two terrific species in Michigan. The Five-lined Skink is widely distributed in the state and is by far our most common lizard. The Six-lined Racerunner is extremely rare in Michigan and is instead a species of special concern.

Lizards are characterized by dry, scaly skin and tails that are often longer than the rest of their bodies. Like most other reptiles, a lizard sheds its skin as it grows. Unlike snakes, a lizard's skin usually splits at the back and comes off in several large sections. Depending upon how fast a lizard grows, it may shed its skin numerous times per year, usually during the summer but also in springtime and fall. Shedding decreases when the animal reaches adult size.

A lizard's tail may detach when grasped by a predator or knocked hard against a stationary object. The separated tail will act as a lure, wriggling for up to several minutes while the lizard escapes. Another tail may regenerate over several weeks, but it will never be as long as the original. The maximum number of times a tail regenerates is unknown.

Lizards are often mistaken for salamanders, which are amphibians. Fundamentally, these are very different animals. Both of our lizard species have dry, scaly skin, clawed toes and external ear openings. Salamanders do not. Lizards and salamanders also live in very different habitats. Lizards tend to be in high and dry environments, while salamanders are rarely far from water.

Occasionally, some people mistake a lizard for a snake. Lizards and snakes are closely related reptiles, but snakes lack legs and are often much larger and longer than lizards. Snakes have unblinking eyes and lack external ear openings. Lizards, on the other hand, have movable eyelids and external ear openings. Lizards usually chew their food before swallowing, while snakes swallow their prey whole.

With a diet of mostly insects, lizards can play a valuable part in controlling pesky bugs such as ants and beetles. They often grab and kill a cricket, beetle or other prey with a quick bite, taking only a few seconds to swallow an insect of average size. At other times they eat their food deliberately, biting to crush and mash the prey before swallowing.

A lizard locates food visually, by scent or through chemical clues sensed with its tongue and probably combines these methods to find prey. Similar to snakes, a lizard will sample the surrounding air with its tongue. Molecules that collect on the tongue are transferred to an organ located in the roof of its mouth (Jacobson's organ), where chemical information is sent to the brain.

Both of our lizard species have powerful jaws and small, strong teeth. Like other reptiles with teeth, lizards have teeth designed for grabbing and holding prey. Most species usually have a row of teeth in each jaw. Our lizards are so small and have such tiny teeth that neither species can deliver a painful bite.

Most species of lizards have excellent eyesight and are able to see in color. Eyes are positioned on the head to allow a maximum view of the surroundings. This may help lizards locate prey while watching for predators such as hawks and foxes.

The males of some lizard species communicate with each other visually by using a series of head and tail movements. They bob their heads while releasing odors from glands and flaunting their often bright-colored bodies, as seen in male Six-lined Racerunners.

For mating, males attack females, biting and riding on top. Lizards have internal fertilization, a process in which the male passes sperm to the female internally, which fertilizes the eggs within her before they are laid.

After mating, the female goes off on her own to deposit up to 6 or more soft, flexible, white-shelled eggs in sandy soil under logs and rocks, where they are incubated by the sun-warmed earth. Michigan lizard eggs are the size of small jellybeans. Unlike some turtles, which have the sex of their offspring determined by nest temperature, the sex of lizard offspring is determined genetically.

Mother skinks are different from the Six-lined Racerunner moms in Michigan. After skinks lay their eggs, they stay around the site. Guarding the eggs and young for several days after hatching, females keep their offspring from being eaten by other lizards and snakes. This might help explain why the Five-lined Skink is such a successful and widespread creature.

Lizard hatchlings look like miniature adults. About 2 inches (5 cm) long upon hatching, they are on their own in days and must fend for themselves. Hatchlings hide beneath rocks and logs for refuge and feed on tiny insects in the dugout. When out and about, they move quickly to avoid being eaten by other lizards and snakes.

Young lizards often have bright blue tails. It is thought that the blue tail is a special diversion since it is easily sacrificed without loss of life and draws a predator's attention away from the vulnerable head. Juveniles grow quickly and become sexually active at 2 years of age. The average life expectancy, however, is thought to be only 5-7 years.

Nearly all lizard species are good diggers and burrowers. Using its slender, pointed snout, a lizard will push its head into sandy soils and start digging in with a swimming motion. Similar to a snake, its body will undulate back and forth as it digs deeper. The digging skill is so good that some species will burrow to below the frost line. There they enter a motionless condition, barely breathing or with a heartbeat for the winter.

Lizards are most active on warm, sunny days. They emerge from underneath logs and rocks early in the day and will bask in the morning light to warm themselves. By midmorning, lizards are warm enough to start looking for food. This is well timed because the insects they eat are also less active when it's cool and are just starting to move around. Lizards retreat under logs and rocks by midday when it becomes too hot. You won't see these animals on cold, rainy days.

older breeding male

juvenile

Five-lined Skink
Eumeces fasciatus

Lizards
(REPTILE)

Family: Skinks (Scincidae)

Size: 5-8" (13-20 cm)

Description: Robust lizard with shiny scales and short legs. Brown to black with 5 narrow lines, gray to pale yellow, on the back. Central stripe splits into a Y with 2 narrow lines across top of head to the nostrils. Older male is often uniformly tan with faint stripes. Breeding male has orange-to-red lips. Juvenile has a bright blue tail.

Eggs/Young: up to 13 eggs once per year; hatchling; 40-60 days; young 2" (5 cm) long and dark with 5 light lines and a metallic blue tail

Origin/Age: native; presumed 5-7 years

Habitat: wet woodlands with rock outcroppings

Overwinter: underground dens below the frost line

Food: crickets, grasshoppers, other insects, snails, spiders

Compare: Look for 5 narrow lines on back with a central stripe splitting into a Y to help identify.

Stan's Notes: This species is considered a woodland lizard, preferring rock outcroppings in woodlands. An excellent digger. Tunnels into soil beneath rocks and logs, where it cools itself on hot summer days. Detaches its tail if grasped or knocked against a fixed object, but can also break off its tail if cornered by pushing it against a rigid surface. Detached tail wiggles around for up to several minutes, presumably to act as a live decoy while the skink makes a quick getaway. Tail will regenerate over time (reportedly 3-4 weeks), but never grows as long as the original. Male and female mature at 3 years. Male finds female by sight and scent. Emerges from dens in early May to bask in the sun. After mating in May, female lays eggs in a chamber she has dug underneath rocks and logs and guards the chamber from predators.

STATUS: Stable 75

Six-lined Racerunner
Cnemidophorus sexlineatus

Lizards
(REPTILE)

Family: Racerunners and Whiptails (Teiidae)

Size: 7-9" (18-22.5 cm)

Description: Male is green on the head and back with blue sides and belly. Six thin light yellow or pale white lines or stripes on the back and sides. Female and juvenile are brown with the same stripes as male. Large hind legs with long toes. Long thin tail is more than half its overall length. Juvenile has a bright light blue tail.

Eggs/Young: up to 6 eggs once per year; hatchling; 50-60 days; young 2" (5 cm) long; appears like adult female

Origin/Age: native; presumed 5-7 years

Habitat: prairies, sandy or gravel soils, hills, rocky bluffs

Overwinter: underground burrows and dens below the frost line

Food: crickets, beetles, other insects, caterpillars, spiders

Compare: Pointed nose and longer legs than Five-lined Skink (pg. 75), which lacks the blue of male Racerunner.

Stan's Notes: The largest lizard in the state and the least cold tolerant. Localized, isolated population. A diurnal lizard. Finds prey by sight and smell. Can chase prey at a speed of up to 18 miles (29 km) per hour for short distances. Seizes prey by mouth. Runs quickly when threatened, hence its common name. Most active at warmest time of day. Doesn't lose its tail as easily as skinks. Drops all or part of its tail when grasped and regenerates a new, shorter tail than the original. Emerges in May from wintering sites on prairies and rocky bluffs. Mates when male is vividly colored in June. Female burrows down to 3 inches (7.5 cm) to lay her eggs. Does not guard eggs. Returns to winter dens in late August.

Salamanders

ENDANGERED
Small-mouthed Salamander

THREATENED
Marbled Salamander

STABLE
Four-toed Salamander
Red-backed Salamander
Eastern Newt
Blue-spotted Salamander
Spotted Salamander
Tiger Salamander
Mudpuppy
Western Lesser Siren

The salamanders include more than 415 species worldwide. Most are concentrated in the Western Hemisphere and ten great species are found in Michigan. Most salamanders are four-legged amphibians with moist skin, long tails and toes without claws, with four toes on the forefeet and five on the hind feet. Western Lesser Sirens have front legs with four toes on each foot, but lack hind legs. Other species, like our Four-toed Salamander, have four toes on both the front and back feet.

Unlike reptiles such as lizards and snakes, salamanders lack scales and have soft, moist skin. Their skin is so sensitive that it can absorb chemicals directly from water, soil or your hands during handling. It is very important to be sure you have clean hands before handling any amphibian.

Salamanders shed their skin from time to time to facilitate growth. This happens in intervals from a few days to several weeks. The old skin often breaks away near the mouth, and the salamander uses its front feet to help extricate itself. Once free, the animal will usually eat its old skin, presumably for the moisture and nutrients. It is rare to see a salamander shed its skin.

Most salamander species have lungs, but nearly all are also able to breathe through their skin. While salamanders take oxygen into their lungs through regular air passages, nearly 80 percent of the respired (released) carbon dioxide is discharged through the skin rather than exhaled. Breathing like this requires a highly specialized skin, blood vessel (vascular) system and red blood cells. The Red-backed Salamander does not have lungs and breathes through its skin. The aquatic Mudpuppy and Western Lesser Siren have large gills on the sides of their heads that extract oxygen directly from the water.

With the exception of the Mudpuppy, the Western Lesser Siren and some Eastern Newts, adult-stage salamanders live on land (terrestrial). In most species, however, egg laying occurs in or close to water. Eggs hatch into carnivorous, aquatic larvae (tadpole-like critters) that breathe with gills. Depending upon water temperature and available food, by late summer or the following spring they transform into air-breathing adults with lungs. At that time they leave the water to live on land. Some species, such as Red-backed Salamanders, live their entire lives terrestrially.

All salamanders need to live in or near moist environments to avoid drying out. They like cool, damp habitats compared to the warm, dry places preferred by lizards and most snakes. Unlike lizards and snakes, salamanders don't bask in the sun and are often most active at night. They are very active during early spring and late fall, when other amphibians are inactive.

Studies show that salamanders can partially freeze their bodies in cold environments and function normally when they thaw. This is possible because of the presence of glycerols, which move from the liver into individual cells and replace the water there. Glycerols lower the freezing point of cells slightly and prevent lethal, cell-rupturing crystals from forming. When the animal thaws, glycerols are replaced with water and cells return to normal function.

A salamander's tail has several functions and is a very important part of its anatomy. Using its tail primarily for locomotion on land, but more importantly in water, a salamander will swing it back and forth in a lateral movement like a fish, which helps propel the animal forward. In some salamander species, the tail is longer and broader in males than females, while in others the tail is used for fat storage and eventually for food. Other salamander males use their tails to waft secretions toward females during mating. Some species of salamanders are able to break off (disjoin) their tails, like lizards, if grabbed or knocked against a rigid surface. A new tail will regenerate over time, but without it a salamander will have less chance of surviving and reproducing.

Salamanders seek and find their food by sight and sense of smell (olfaction). They rely primarily on their motion-sensitive eyes, color vision and, to a lesser extent, scent for locating food in dark places or underwater.

All salamanders have teeth. Similar to snake teeth, salamander teeth are curved toward the back of the mouth. This formation helps them grab and hold prey and swallow. Teeth are shed and replaced throughout the life of a salamander. Salamanders have one to two rows of teeth in both the upper and lower jaws. The jaws are not very powerful, so even a large salamander cannot hurt you if it bites.

Salamanders swallow their food with the additional help of their eyeballs. That's right–their eyeballs! Eyeballs in these critters are hard and eye sockets extend into the mouth. When a salamander swallows, its eyeballs sink into the sockets and help crush food and force it into the throat.

Like frogs and toads, salamanders have long tongues that extend well beyond their mouths. Their tongues attach to the lower front part of the mouth instead of at the back, where attachment occurs in most other animals. When a salamander sees a cricket or other prey that is not too fast or large, it will lean forward on its toes and wait. Once the prey moves closer, the salamander will quickly thrust forward and simultaneously roll out its tongue to make contact. A sticky substance on the tip of the tongue holds the prey fast while it is drawn into the mouth. This all happens in 10 milliseconds–slower than in most frog and toad species, but faster than the human eye can see. A mucous-producing gland in the tongue pad and an adhesive-secreting gland on the roof of the mouth work together to replenish the sticky substance on the end of the tongue.

If the prey is long like an earthworm, the salamander will hold it in its mouth and shake its head quickly back and forth. With each bite it will toss the worm farther back into its mouth and down the throat. Smaller food is swallowed quickly, in just a few seconds, with few bites and no shaking.

It is believed that salamanders can hear only very loud sounds. They lack external ear openings and middle ear cavities, but they have functioning inner ears that permit some reception of sound. Generally silent, the Western Lesser Siren and some other species have the ability to make faint clicking or hissing noises, possibly produced by opening or closing the nasal valves or passageway to the throat.

Unfortunately, due to several factors, populations of salamanders in Michigan have been decreasing. Loss of habitat, chemicals in the environment and increased ultraviolet light are thought to be the main causes for the decline.

Four-toed Salamander
Hemidactylium scutatum

Salamanders
(AMPHIBIAN)

Family: Lungless Salamanders (Plethodontidae)

Size: 2-3" (5-7.5 cm)

Description: An overall reddish brown body with irregular silver spots or flecks on sides with 13-14 costal grooves on each side, and V-shaped grooves on the back. White belly (venter) with dark spots. Constricted at base of tail. Has 4 toes on all feet.

Eggs/Young: more than 20 eggs once per year; larva; hatches into larva in 30-60 days, transforms into adult in 3-8 weeks, in late summer; the warmer the water, the less time needed to transform

Origin/Age: native; 10 years

Habitat: bogs in moist coniferous woods, often near springs

Overwinter: under rocks, logs, leaves or underground in burrows

Food: small insects

Compare: Red-backed Salamander (pg. 85) is larger and has a bright red back. Unlike most terrestrial salamanders, the Four-toed has 4 toes instead of 5 on its hind feet.

tan's Notes: Our smallest salamander, named for the number of toes n hind feet. Very secretive, often hiding in thick mats of sphagnum noss. Difficult to see due to its colors and residence in inaccessible oggy habitats. Sometimes active after heavy rains at night in spring, vhen it moves to breeding sites. Emerges in April to mate. Female ays eggs (oviparous) in sphagnum moss over water, attaching each gg singly to the underside of clumps of moss. Guards them until the arvae emerge and drop into water below. After young transform, they nove back onto land to spend the winter with adults. Tail breaks off t the base if grabbed. Regenerates tail.

dark back

juvenile

Red-backed Salamander
Plethodon cinereus

Salamanders
(AMPHIBIAN)

Family: Lungless Salamanders (Plethodontidae)

Size: 2-4" (5-10 cm)

Description: Thin salamander with a bright red back. Some have backs that are dark red to nearly brown (see inset). Sides and belly are light brown to gray with white flecks, with 18-20 costal grooves per side. Delicate thin legs. Large bulging eyes. Has 5 toes on hind feet.

Eggs/Young: up to 14 eggs in a cluster every other year; hatchling; hatches into "miniature adult" in 30-60 days

Origin/Age: native; presumed 5-7 years

Habitat: deciduous and coniferous forests

Overwinter: underground burrows from several inches to 3 feet (.9 m) deep

Food: worms, ants, other small insects, centipedes, spiders

Compare: Seemingly more abundant than Four-toed (pg. 83), which has 4 toes on its hind feet.

tan's Notes: One of our smallest salamanders. Completely terrestrial, t is one of two Michigan salamanders that doesn't lay eggs in water. Over 230 Lungless species, with only two in Michigan. Common in ortheastern U.S. Has a home range of 270 square feet (24 sq. m). Frequently found in a small area in moist woodlands with downed branches and bark. Lacks lungs, breathing through its skin. Seen in moist habitats because it needs to keep its skin moist to breathe. Hunts insects during days and nights, especially on rainy nights. pends days under leaves, rocks and logs. Matures at 2 years. Female ays eggs on the undersurface of a rock or log and guards them for up to 3 weeks after hatching. Young hatch looking like miniature dults (see inset) unlike other salamanders, which first go through an quatic larva stage.

aquatic adult

terrestrial adult

eft

Eastern Newt
Notophthalmus viridescens

Salamanders
(AMPHIBIAN)

Family: Newts (Salamandridae)

Size: 2-5" (5-13 cm)

Description: Olive green back with a row of red spots on the sides. Lacks costal grooves. Yellow belly with black spots. Dark line through the eyes. Tail is vertically flattened to form a fin. Juvenile (eft) is terrestrial and smaller than aquatic adult, with dry, granular, dull orange-to-brown skin.

Eggs/Young: 10-100 eggs several times per year; larva; hatches into larva in 14-26 days, transforms into terrestrial juvenile (eft) in 2-5 weeks, in late summer and stays juvenile for 2-3 years; sometimes skips the eft stage and transforms into aquatic adult; the warmer the water, the less time needed to hatch and transform

Origin/Age: native; 10 years

Habitat: ponds, lakes, streams, wetlands in or next to forests

Overwinter: adult remains active in deeper lakes, terrestrial adult and eft spend winters under leaves, logs and rocks

Food: aquatic insects, earthworms, fish eggs, fairy shrimp

Compare: Mudpuppy (pg. 99) and Western Siren (pg. 101) have red external gills. Terrestrial juvenile (eft) has dry, granular skin and is frequently darker than the Red-backed Salamander (pg. 85).

Stan's Notes: A unique development. Eggs hatch into aquatic larvae, which transform into terrestrial juveniles (efts). Some skip the eft stage and transform into aquatic adults. Aquatic adults are active in streams and ponds year-round. Some transform into terrestrial adults, appearing dark green or brown on top with orange yellow bellies. Adults and efts secrete a toxin to repel predators. Eats mosquito larvae.

Blue-spotted Salamander
Ambystoma laterale

Salamanders
(AMPHIBIAN)

Family: Mole Salamanders (Ambystomatidae)

Size: 3-5" (7.5-13 cm)

Description: Back (dorsum) is dark brown to nearly black with a lighter, often gray belly (venter). Irregularly shaped and distributed blue flecks or spots, concentrated mostly on tail and sides, with 13 costal grooves on each side (lateral). Juvenile has yellow speckles.

Eggs/Young: up to 500 eggs in several clusters once a year; larva; hatches into larva in 3-5 weeks, transforms in 50-60 days; the warmer the water, the less time needed to hatch and transform; larva has dark mottling or spots on tail

Origin/Age: native; 5-10 years

Habitat: moist deciduous and mixed coniferous forests

Overwinter: under rocks and logs at edges of ponds

Food: insects, earthworms, spiders, slugs, centipedes

Compare: Small-mouthed Salamander (pg. 93) has a smaller head, white spots and is less common.

Stan's Notes: This is a very cold-tolerant salamander, most common in woodlands with permanent ponds. Usually is seen only during spring and fall movements and migrations. Can be found in summer beneath logs in moist habitats, similar to Red-backed Salamanders. Thrashes about when grabbed or attacked by a predator and releases a foul-tasting, sticky substance. Female is slightly larger than male. Male has a relatively longer, more flattened tail. Moves to breeding ponds in April. Mates shortly after arriving. Mating occurs mostly at night and underwater, where male clasps female while rubbing his nose over her body. After mating, the female lays eggs that attach to submerged aquatic vegetation.

STATUS: Stable   **89**

Marbled Salamander

Ambystoma opacum

Salamanders
(AMPHIBIAN)

Family: Mole Salamanders (Ambystomatidae)

Size: 3-5" (7.5-13 cm)

Description: Medium salamander with a wide body and thick tail. Overall gray to nearly black with bold white (sometimes silver) marks, often forming bands across body. Black belly. Wide head. Large protruding eyes. Short stout legs. Male often brighter white than female.

Eggs/Young: 50-200 eggs in loose clusters once per year; larva; hatches into larva in 14-24 days, transforms into terrestrial adult in 30-45 days

Origin/Age: native; 7-9 years

Habitat: moist lowland forests, river valleys

Overwinter: underground mammal dens and tunnels, burrowing into soil down to 2 feet (.6 m)

Food: insects, earthworms, slugs, snails, insect larvae

Compare: Smaller than the Tiger Salamander (pg. 97), which is black with yellow markings. The Small-mouthed Salamander (pg. 93) lacks cross bands and is much thinner with smaller white marks.

Stan's Notes: Considered common in many areas of the U.S., but has a very restricted range in Michigan. Spends much time beneath logs and rocks, burrowing deeper into soil during drought. Usually only seen during the autumn breeding season, when adults move around on warm, moist or rainy nights. After mating, female lays eggs in a shallow depression or cavity beneath fallen logs or moss. She often curls around her eggs, waiting for rain to fill the depression, after which the eggs are on their own. Larvae disperse when spring rains flood the nest. Young often feed upon Tiger Salamander larvae and transform into terrestrial adults in late spring or early summer.

STATUS: Threatened   91

Small-mouthed Salamander
Ambystoma texanum

Salamanders
(AMPHIBIAN)

Family: Mole Salamanders (Ambystomatidae)

Size: 4-7" (10-18 cm)

Description: Long thin salamander with a tiny head in proportion to body. Light gray to nearly black with dull white markings (sometimes unmarked). Dark belly (sometimes with white flecks). Has 14-16 costal grooves.

Eggs/Young: 300-700 eggs in small, gelatinous clusters attached to sticks or leaves once per year; larva; 3-8 weeks depending upon the water temperature, transforms into 2" (5 cm) long terrestrial adult in 30-45 days

Origin/Age: native; 7-9 years

Habitat: moist forests, river valleys, tall grass, along streams

Overwinter: underground dens and tunnels, crayfish burrows

Food: insects, worms, slugs, snails, other salamander larvae

Compare: Tiger Salamander (pg. 97) has a large, wide head and yellow markings. Longer and thinner than Marbled Salamander (pg. 91), which has bold white marks and a large, wide head.

Stan's Notes: At northern limits of range in Michigan. One of the first salamanders seen in spring in shallow, warm woodland pools. Large groups move toward traditional breeding ponds when spring rains begin. Gathering in courting groups, males will walk around females, nudge them with their blunt snouts and leave sperm packets on the ground. After taking up a packet into the cloaca, the females lay eggs. Like the other mole salamanders, this species has concentrations of granular glands on the upper surface of its tail that produce toxic secretions to repel predators. If threatened, it will lift and wave its tail while hiding its head. Male is often smaller than female, with a longer tail. Hybridizes with Blue-spotted Salamander where ranges overlap.

STATUS: Endangered 93

Spotted Salamander
Ambystoma maculatum

Salamanders
(AMPHIBIAN)

Family: Mole Salamanders (Ambystomatidae)

Size: 4-9" (10-22.5 cm)

Description: Dark brown to black with 2 rows of irregular-spaced yellow spots that run from head to tail. May also have orange (rarely white or tan) spots, especially near the head. Sides are often lighter than the back, with 11-13 costal grooves on each side. Unspotted gray belly. May have spots on legs. Wide head and round snout.

Eggs/Young: up to 250 eggs in a very stiff, globular mass once per year; larva; hatches into larva in 20-50 days, transforms into an adult in 40-50 days, in August and September; the warmer the water, the less time needed to hatch and transform

Origin/Age: native; presumed 20-25 years

Habitat: moist forests, woodland ponds

Overwinter: underground burrows near breeding ponds

Food: insects, earthworms, spiders, slugs

Compare: Tiger Salamander (pg. 97) has irregularly shaped yellow blotches and lacks a gray belly. The Marbled Salamander (pg. 91) has large white or silver marks.

tan's Notes: Collected for the pet trade, resulting in greatly lowered populations in some regions of the eastern U.S. Lowers its head and waves tail if approached by a predator. Releases a poison from glands at the back of its head and on tail to repel predators. Uses an underground burrow during summer days; sometimes comes out at night. An active breeding season in spring and less conspicuous afterward. Migrates to same breeding ponds in April and May each year, with males arriving before females. Male matures at 2 years, female at 3-5 years. Eggs attach to submerged vegetation on bottom of ponds.

STATUS: Stable 95

Tiger Salamander

Ambystoma tigrinum

Family: Mole Salamanders (Ambystomatidae)

Size: 7-13" (18-33 cm)

Description: Back (dorsum) is dark brown to nearly black. Each side has 11-14 costal grooves. Dull yellow blotches or spots, irregularly spaced. Many get more blotchy with age. Some are all black or nearly all black. Short stout legs. Broad head, round snout and small eyes.

Eggs/Young: up to 100 eggs in a large round mass once per year; larva; hatches into larva in 40-45 days, transforms into terrestrial adult in 4-6 months; some larvae will overwinter and transform in spring; the warmer the water, the less time needed to hatch and transform

Origin/Age: native; 10-25 years

Habitat: prairie potholes, ponds, lakes, woodlands

Overwinter: underground mammal dens and tunnels, burrowing into soil down to 2 feet (.6 m)

Food: earthworms, insects

Compare: Spotted Salamander (pg. 95) has brighter yellow (sometimes orange) spots near head and a gray belly. Marbled Salamander (pg. 91) is less common than Tiger Salamander and has white to silver markings.

Stan's Notes: The largest land salamander in the state. An aggressive predator, called Tiger due to its colors. Many get caught in window wells, where they die if not removed. Most adults spend their time in burrows. Usually seen crossing roads during migration to and from breeding ponds. Breeds in its first spring. Lays eggs on plants near pond bottom. Most Tiger Salamanders seen in fall are juveniles. Misnamed Mudpuppy, which is a different species. Like Mudpuppy,

STATUS: Stable **97**

external gills

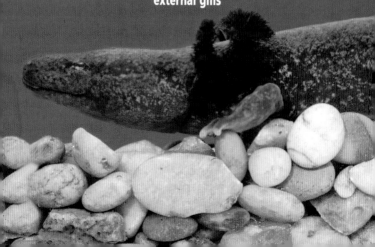

Mudpuppy
Necterus maculosus

Salamanders
(AMPHIBIAN)

Family: Mudpuppies (Proteidae)

Size: 10-16" (25-40 cm)

Description: An overall dark brown salamander with dark spots or blotches. Deep red external gills at base of head. Feet have 4 toes. Short legs. Tail is vertically flattened to form a fin. Squared head and tiny eyes. Juvenile is dark with some yellow stripes on each side (lateral).

Eggs/Young: up to 100 eggs once per year, in late spring; larva; hatches into larva in 30-60 days, transforms in 40-60 days; the warmer the water, the less time needed to hatch and transform

Origin/Age: native; 15-25 years

Habitat: lakes, rivers, reservoirs, rocky streams

Overwinter: remains active year-round underwater

Food: aquatic insects, crayfish, worms, fish, salamanders

Compare: The Eastern Newt (pg. 87) and Western Lesser Siren (pg. 101) are also aquatic salamanders, but Newt is smaller and lacks red gills, and Siren lacks hind legs.

tan's Notes: One of our largest salamanders. All-aquatic with lungs nd gills. Individuals that are in well-oxygenated rivers have shorter ills. Adults stay in deeper water, young are in shallow water. Mostly octurnal, feeding at night. Stays active in same territory year-round. refers rocks and logs for cover. Sometimes caught by fishermen and old in bait shops as Waterdog. Four toes on each foot. This is unlike most terrestrial salamanders, which have five toes on hind feet. Breeds n fall to early winter. Lays eggs on roof of an underwater cavity that usually faces downstream, below rocks and logs. Eggs are suspended rom cavity roof by a gelatinous stalk. Female guards eggs until they atch. Only six species worldwide in this family, five in the U.S.

external gills

Western Lesser Siren
Siren intermedia

Salamanders
(AMPHIBIAN)

Family: Siren (Sirenidae)

Size: 7-20" (18-50 cm)

Description: Aquatic salamander with an eel-like body, no hind legs and vertically flattened tail. Tiny front legs with 4 toes on each foot. Bushy gray and red gills. Overall light gray to brown with many dark marks (sometimes uniformly black). Gray belly with tiny light flecks. Tiny eyes. Has 34-35 costal grooves. Male can be larger, but often no difference between sexes.

Eggs/Young: 200-500 eggs in small clusters in a protected cavity or depression once a year; larva; unknown number of days to hatch and transform

Origin/Age: native; 20 or more years

Habitat: ponds, slow rivers, shallow lakes, mud-bottomed lakes with abundant vegetation

Overwinter: underwater, burrowed into mud or active all winter

Food: aquatic insects and plants, small crustaceans, snails

Compare: Mudpuppy (pg. 99) is similar in size, but has 4 legs and lacks an eel-like body.

tan's Notes: Very rare, all-aquatic, eel-like salamander with external ills and no hind feet. This is the only salamander in the state to make ound, hissing, squeaking and making clicking noises when handled. More active at night, hiding in warm pond bottoms during the day. Moves overland, retreats into crayfish tunnels or burrows into mud when ponds dry. Produces a protective cocoon that covers the body xcept for the mouth to survive severe drought. Matures in 2-3 years. an live more than 20 years. Not much more is known about this ecretive salamander. Mainly in the South, reaching its northernmost imits in Michigan, with three subspecies in North America.

Frogs and Toads

SPECIAL CONCERN
Cricket Frog

STABLE
Western Chorus Frog
Spring Peeper
Cope's Gray Treefrog
Gray Treefrog
Wood Frog
Pickerel Frog
Mink Frog
Northern Leopard Frog
Green Frog
Bullfrog
Fowler's Toad
American Toad

Frogs and toads were among the first animals with backbones to evolve. Fossil evidence shows frog-like creatures dating back to more than 200 million years, just before the majority of dinosaurs. Found on every continent except Antarctica, there are over 4,000 frog and toad species worldwide. All except one species lack body scales and toenails and are tailless as adults. All have large eyes and very sensitive skin. Outwardly similar, there is also no scientific separation between frogs and toads. However, while toads are considered a type of frog, frogs are not considered a type of toad.

Frogs live mostly in water (aquatic) and toads live mostly on land (terrestrial). Frogs usually have smooth, moist skin. Toads have dry, warty skin. Their skin is also thicker, reducing moisture loss and allowing them to live farther away from water than frogs. Frogs have more webbing between their toes than toads and more powerful legs, which propel them great distances when hopping or swimming. Toads have much weaker hind legs that restrict their hopping to only a few inches at a time, but permit strong swimming. Most frog and toad species are active after dark (nocturnal). Some are active during the day (diurnal).

Adult frogs and toads are carnivorous amphibians, well equipped for finding and capturing food. Most will feed on insects, sighting prey easily with their large, bulging eyes. Most have excellent color vision, but rely on movement to focus on their prey. Some species also use their sense of smell (olfaction), to a lesser extent, to detect food, usually in dark environments.

When a potential insect meal is spotted, a frog may slowly move closer to it without striking. If the bug freezes and remains still, the frog may eventually leave to find another insect. If the bug moves, the frog will instantly unfurl its long tongue. Like many other amphibians, the tongues of frogs and toads attach at the front of the mouth and have a sticky pad on the tip. The sticky pad adheres to the prey and allows it to be snatched into the mouth. However, if you watch a frog capture prey, you won't be able to see the tongue because it moves out and in faster than the human eye can see.

Most frogs and toads have one row (sometimes two) of teeth in the upper jaw and lack teeth in the lower jaw. Some species have teeth in the center of the roof of the mouth that help them hold wiggling prey. If a meal is too large and dangles from the mouth, a frog or toad will sometime use its front feet to stuff the morsel down its throat. It only takes a few seconds for it to swallow a food item as large as a cricket.

Like salamanders, the eyeballs of frogs and toads help them eat. Their large, hard eyeballs are set in sockets that lack an underlying bony structure. When a frog swallows, its eyes sink into the mouth cavity. The bottom of the eyeballs help mash and push food into the throat.

One study of vision in frogs and toads has shown that frogs prefer the color blue, while toads prefer green. In experiments, frogs jumped toward a blue light. Since frogs will jump into water as a survival response, it makes sense that they have an attraction to blue, the color of most water. Toads, on the other hand, jumped toward a green light. Green is representative of vegetation, where toads prefer to hide.

The skin of frogs and toads performs many functions. It protects against disease and trauma, aids in breathing and absorbs and releases extra water. Some frogs and toads have skin that changes color depending upon environmental factors such as temperature or surroundings. In many species, granular glands underneath the skin enable frogs and toads to excrete toxic chemicals as a defense system. These glands are the bumps or "warts" often seen just behind the eyes. They secrete a milky yellowish, sometimes colorless, often sticky substance that is very distasteful to many animals and can cause burning and irritation in the eyes, nose and mouth. It may also cause gastrointestinal upset and vomiting if swallowed.

Like salamanders, all frogs and toads shed their skin from time to time as they grow. The shed can also be called the molt and occurs in intervals from a few days to a couple weeks. Molting starts with the old skin splitting down the back. After pulling its legs out of its skin, a frog will work the remaining skin off its body and toward the mouth, where it is swallowed. During this process it appears as though the critter is writhing in pain or going through a convulsion. Once the skin is removed and eaten, the animal returns to its normal conduct.

Most frogs and toads have well-developed ears and hearing. They have eardrums, sound-transmitting bones in middle ear cavities and nerve endings in their inner ears to receive and transmit sound to the brain. The eardrums are often large and obvious, appearing as round, flat disks on the side of the head just behind the eyes. In some species, such as the Bullfrog, males have larger eardrums than females. It is thought that ears are used not only for hearing the calls of males, but also to detect sounds made by storms and rainfall, triggering the emergence of some species from underground retreats in spring. A frog's hearing is so good it can differentiate among several species of frogs calling together.

Frogs and toads are the only amphibians that produce organized sound. Males give advertisement calls to attract females for mating during spring and summer. Both males and females can produce distress or warning calls, and release calls if grabbed by a predator or another frog. For more information about calls, see the Calls of Frogs and Toads section and the Croaking Chart on pages xviii-xxiii.

When a male's call successfully attracts a female, they will mate. Frogs and toads breed in water. The male holds the female in a sexual grasp called amplexus, wrapping his front legs around her and clasping her with his enlarged thumbs. Roughened pads on his front feet also help him keep a firm grip. As the female lays eggs (oviparous), the male releases sperm onto them. This process is called external fertilization. Afterward, the female may give a vibrating release call, signaling the male to let her go.

Depending upon the species, a mass of frog eggs appears as a clear, gelatinous round ball measuring 2-4 inches (5-10 cm) in diameter. The developing embryo appears as a dark center within each clear egg and darkens the entire egg as it reaches maturity and nears hatching. Frog eggs are laid in water, often attaching to aquatic vegetation or floating freely. Toad eggs are also laid in the water, sometimes attaching to plants or floating freely. They are usually clear, gelatinous strings as long as 12 inches (30 cm).

Water, oxygen and carbon dioxide flow back and forth through the egg membranes, sustaining the embryos during development into aquatic larvae (tadpoles). The temperature of water influences how long the eggs will take to hatch. The warmer the water, the faster the eggs develop and hatch into tadpoles, and the faster the tadpoles grow and transform into frogs or toads. The reverse is true of cold water.

Females of some frog species, such as Wood Frogs, lay their egg masses with other females in small bodies of water, where eggs cluster into one large group. Studies show that the larger clusters absorb more sunlight, which raises the temperature within the egg mass to as much as 2°F (-17°C) above the temperature of the surrounding water. These warmed eggs hatch faster, resulting in less exposure to predators.

In some species, tadpoles can be seen in large groups known as schools. They move around like a flock of birds, twisting and turning in water, more like one large organism rather than hundreds of individuals. A tadpole has as many as three pairs of internal gills for breathing, a round head and long, thin, fin-like tail. Most have tiny, replaceable, teeth-like projections (denticles) for biting or scraping their food. They usually feed on algae along with other aquatic plants (herbivores) until they start to transform (metamorphose) and grow legs. Tadpoles do not eat during late metamorphosis, but use stored fats in the tail, causing the tail to absorb into the body. For reasons unknown, many die at this stage. Most that do survive transform into frogs and toads before summer's end, while some species overwinter in the tadpole form.

Frogs and toads routinely migrate to the same breeding sites each spring. Like salmon, they return to their place of birth for mating. Returning to the birthplace makes sense because it has already proved itself as a good location to reproduce. Males usually arrive before females and start giving their advertisement calls. Females arrive a couple days to a week later. After mating, they frequently journey to familiar areas where there is abundant food. In autumn they move again, this time to places where they will overwinter. Most species travel as far as a couple hundred yards. Some will migrate up to several miles.

Their ability of homing to specific breeding ponds has been well studied. The memory of a breeding location is so deeply rooted that individual frogs and toads of some species keep returning to familiar places even when their breeding ponds were destroyed years earlier. Interestingly, they follow the same path each year, entering and leaving the habitat at the same spot, usually within a few feet.

Most frogs and toads migrate at night, especially when humidity is high or during rain. They may be using the setting and rising sun to orient themselves and the moon to guide their movements. Similar to birds, they sense light through a photoreceptive organ in the skull known as the pineal organ and, to a lesser extent, with their eyes. Frogs and toads also use their sense of smell to find their way home or to favorite breeding ponds. Olfaction, however, is most likely used in conjunction with celestial orientation and not on its own. Other sensory receptors, such as familiar landmarks, may also be at work to help guide their movements.

All of our frog and toad species overwinter in a motionless state. Western Chorus Frogs, Gray Treefrogs and others that overwinter on land freeze 40-70 percent or more of their bodies–a condition thought impossible not too long ago–to survive the freezing and bitter cold of winter. Species such as Western Chorus Frogs, Gray Treefrogs and others simply get underneath leaf litter or a rock or log. The rest of our frogs overwinter underwater, burrowing into or resting on the bottom of a lake or pond, presumably without freezing or moving. Toads burrow deep into the earth to the frost line or just below to wait out the season.

Regrettably, populations of frogs and toads worldwide are on the decline. Many of our species, such as Northern Leopard Frogs and Cricket Frogs, are in serious decline and have disappeared from specific locations throughout Michigan. There are probably many reasons for this decline and the subject is being closely examined. Because amphibians such as frogs and toads are useful in measuring the effects of changes in the environment, many people consider them to be "bio-indicators." To help protect the future of these valuable creatures, treat all of our species with respect.

green morph

brown morph

Western Chorus Frog
Pseudacris triseriata

Frogs and Toads
(AMPHIBIAN)

Family: Treefrogs (Hylidae)

Size: ¾-1" (2-2.5 cm)

Description: A slender frog, tan to brown (sometimes light brown to red, gray and green) with 3 darker stripes on the back. Belly is tan to cream and unmarked. Distinctive white upper lip. Dark stripe from snout through eyes and down sides. Male has a dark throat (vocal) sac.

Eggs/Young: 500-2,500 eggs in small round clusters of 20-300 eggs once per year; tadpole; hatches into tadpole in 3-14 days, transforms in 50-70 days; the warmer the water, the less time needed to hatch and transform

Origin/Age: native; 3-5 years

Habitat: wetlands, moist woodlands, wet meadows and fields

Overwinter: under leaf litter, logs and rocks; partially freezes

Food: flies, crickets, beetles and other small insects, spiders

Compare: Peeper (pg. 113) has a distinctive X marking on the back. Cricket Frog (pg. 111) has a triangular mark between eyes and bumpy skin.

Stan's Notes: Widespread in the Lower Peninsula; seen in scattered locations in the Upper Peninsula. One of the earliest breeders, with one of the longest calling seasons. Call sounds like a thumb running down the teeth of a stiff pocket comb. The warmer the water, the faster the call. Sings in large groups (choruses), hence the common name. Not a good climber. Male is smaller than female. After mating, becomes secretive, staying near ponds, feeding mostly during the morning and evening. Lays eggs on submerged plants. Temporary breeding ponds may dry up before tadpoles transform, leading to reproductive failure. Also called Striped Chorus Frog. A subspecies known as Boreal Chorus Frog is found only on Isle Royale.

green morph

bronze morph

Cricket Frog
Acris crepitans

Frogs and Toads
(AMPHIBIAN)

Family: Treefrogs (Hylidae)

Size: ½-1½" (1-4 cm)

Description: Overall brown to gray or green, sometimes with a green or bronze stripe on middle of back. Cream or white belly, unmarked. May have a dark green or bronze triangular patch between eyes. Textured skin with tiny dark bumps. Some have no markings.

Eggs/Young: 200-400 eggs in small round masses or clusters once per year; tadpole; hatches into tadpole in 3-7 days, transforms in 35-70 days; the warmer the water, the less time needed to hatch and transform

Origin/Age: native; 2-5 years

Habitat: along streams, ponds, lakes and rivers

Overwinter: under leaf litter, logs and rocks near water; presumed to partially freeze

Food: small aquatic insects

Compare: Western Chorus Frog (pg. 109) has dark stripes and white upper lip. Peeper (pg. 113) has an X marking on back. Use location to help identify this rare frog.

an's Notes: Variable markings from green to bronze. Population has eclined in the state since the 1970s. Was seen throughout southern ichigan; now found in a few select areas. Michigan is near the north-n edge of its range in the U.S. Usually seen along more open edges permanent ponds, lakes and slower rivers. Can jump up to 4 feet .2 m) to escape predators. One of the last frogs to mate (late May to ly). Males sing choruses of a metallic "glick-glick-glick" (like two etal balls clicking together) during days or evenings, especially after in. Song lasts up to 30 seconds. Eggs are laid underwater and attach aquatic vegetation. Overwintering starts in August.

STATUS: Special Concern **111**

Spring Peeper
Pseudacris crucifer

Frogs and Toads
(AMPHIBIAN)

Family: Treefrogs (Hylidae)

Size: ¾-1¼" (2-3 cm)

Description: Overall light brown, but can be dark brown to gray with many dark markings. A distinctive X-shaped marking on the back (dorsum). Belly is tan to cream and unmarked. Male has a brown throat (vocal) sac.

Eggs/Young: 750-1,300 eggs in small groups of up to 100 once per year; tadpole; hatches into tadpole in 4-10 days, transforms in 45-90 days; the warmer the water, the less time needed to hatch and transform

Origin/Age: native; 3-5 years

Habitat: woodland ponds, wetlands, wet meadows

Overwinter: under leaf litter, logs and rocks; partially freezes

Food: small aquatic insects, terrestrial insects such as flies

Compare: Similar size as Chorus Frog (pg. 109), but lacks the dark line through eyes and dark stripes on back.

tan's Notes: One of the first frogs to call in spring. Breeds early in the ear to reduce exposure to predators. Call sounds like high-pitched iny bells ringing (one each second) or peeping, hence the common ame. Some males don't call during breeding season. Silent (satellite) nales stay near a calling male and try to intercept the females that are ttracted to the caller. Females prefer the louder, faster-calling males, vhich tend to be older and larger. Calls may be heard in late summer o fall. Eggs are laid in water and attach to submerged plants. Moves nto woodlands after breeding to feed on insects. Color changes with emperature, colder temperatures resulting in darker frogs. Large X nark on its back is variable in color and completeness, and accounts or the Latin species name *crucifer*, meaning "cross bearer."

green

belly

gray

Cope's Gray Treefrog
Hyla chrysoscelis

Family: Treefrogs (Hylidae)

Size: 1¼-2¼" (3-5.5 cm)

Description: A green frog with an unmarked white belly. Changes color to gray with large dark spots and blotches. Some have a small white patch just below each eye. Dark stripe extending from behind eyes to front legs and beyond. Inside thighs of hind legs are yellow. Large toe pads. Male has a gray throat (vocal) sac.

Eggs/Young: up to 2,000 eggs in small clusters of 10-50 eggs once per year; tadpole; hatches into tadpole in 3-7 days, transforms in 42-56 days; the warmer the water, the less time needed to hatch and transform

Origin/Age: native; 5-7 years

Habitat: moist forests, woodland edges, ponds, wet meadows

Overwinter: under leaf litter, logs and rocks; partially freezes

Food: beetles and other insects, caterpillars

Compare: The Gray Treefrog (pg. 117) is nearly identical. The best way to differentiate them is by their calls (audio CD tracks 20-29). Western Chorus Frog (pg. 109) and Spring Peeper (pg. 113) lack enlarged toe pads.

Stan's Notes: Color changes with temperature, humidity or habitat, in seconds or up to an hour. Has mucus-producing glands on toe pads. Adheres to varied surfaces such as tree bark and glass. Often seen on windows after dark, hunting bugs. Male is smaller than female. Males call in loud choruses from ponds and trees. Sings a fast, metallic trill lasting 1-3 seconds. Silent (satellite) males wait near a male caller to intercept females for mating. Moves to trees after breeding season to feed on insects. Found up to 30 feet (9 m) off the ground. Treefrogs family is large, with more than 700 species worldwide.

Gray Treefrog
Hyla versicolor

Family: Treefrogs (Hylidae)

Size: 1¼-2½" (3-6 cm)

Description: Green frog with a white belly. Changes color to gray with dark spots and blotches outlined with a black line. Some have a small white patch below each eye. Dark stripe extending from behind eyes to front legs and beyond. Inside thighs of hind legs are yellow. Large toe pads. Male has a gray throat (vocal) sac.

Eggs/Young: up to 2,000 eggs in small clusters of 10-50 eggs once per year; tadpole; hatches into tadpole in 3-7 days, transforms in 42-56 days; the warmer the water, the less time needed to hatch and transform

Origin/Age: native; 5-7 years

Habitat: moist forests, woodland edges and ponds

Overwinter: under leaf litter, logs and rocks; partially freezes

Food: beetles and other insects, caterpillars

Compare: Nearly identical to Cope's Gray Treefrog (pg. 115). Listen to calls to identify (audio CD tracks 20-29).

Stan's Notes: Considered the same species as the Cope's Gray Treefrog until 1968. Changes color with the temperature, humidity or habitat. Produces a sticky substance that allows it to climb smooth surfaces such as glass without trouble. Freezes 40-70 percent or more of its body when overwintering, replacing water in vital organ cells with glycerols. Glycerols don't form the pointed ice crystals produced by water that rupture cell walls. Ice still forms in the body, but not in the vital organs. Male is smaller than female. Males sing a slower trill than Cope's that lasts 1-3 seconds. A large chorus can be very loud. Young frogs leave ponds in July and August to join adults on land. Much more common than the Cope's Gray Treefrog.

yellow morph

red morph

brown morph

Wood Frog
Rana sylvatica

Frogs and Toads
(AMPHIBIAN)

Family: True Frogs (Ranidae)

Size: 1-3" (2.5-7.5 cm)

Description: Variable in color, usually light tan to brown. Colors of other individuals range from pale yellow to deep red (see insets). Characteristic dark face mask from snout to just behind tympanums (eardrums). White belly, occasionally with dark mottling. White upper lip. Dark lines along prominent dorsolateral folds.

Eggs/Young: up to 3,000 eggs in gelatinous, floating masses once per year; tadpole; hatches into tadpole in 4-14 days, transforms in 42-100 days; the warmer the water, the less time needed to hatch and transform

Origin/Age: native; 3-5 years

Habitat: ponds in forests and prairies, wetlands, moist woods

Overwinter: under leaf litter, logs and rocks; partially freezes

Food: flying insects, beetles, crickets

Compare: Western Chorus Frog (pg. 109) shares a white upper lip but lacks prominent dorsolateral folds. The dark face mask of Wood Frog is a unique field mark.

Stan's Notes: Common woodland pond frog. Adults are seen in large numbers crossing roads in spring and fall when moving to and from water and woodlands. Juveniles are seen along edges of water in late summer. Males call a low, duck-like quack or chuckle in choruses from surface of water. Adult male is smaller than female and has larger thumbs on forefeet. Female often deposits eggs with other females, producing huge numbers of eggs in one spot. After breeding, moves to forests for the remainder of summer while tadpoles transform. The northernmost frog species, found in Canada and Alaska.

Pickerel Frog
Rana palustris

Frogs and Toads
(AMPHIBIAN)

Family: True Frogs (Ranidae)

Size: 1½-3" (4-7.5 cm)

Description: Overall tan with 2 rows of large, squarish dark spots on back between dorsolateral folds. Spots on sides. White belly (venter) and throat. Light line on upper lip. Distinct dorsolateral folds along the length of the body. Inside thighs of hind legs are bright yellow.

Eggs/Young: 500-3,000 eggs in several round clusters once per year; tadpole; hatches into tadpole in 10-21 days, transforms in 60-90 days; the warmer the water, the less time needed to hatch and transform

Origin/Age: native; 5-7 years

Habitat: wetlands, ponds, cold streams, woodlands

Overwinter: underwater; burrows into mud on the soft bottom of ponds and streams

Food: grasshoppers, beetles and other insects, slugs, snails

Compare: Northern Leopard Frog (pg. 125) is larger and has green inner thighs.

Stan's Notes: Not common in Michigan, but can be locally abundant. Prefers cold water streams and avoids stagnant, isolated warm water; however, female often lays her eggs in warmer water to increase the development rate. Avoid collecting, harming or moving this species. Emerges in April. Male gives a soft, low-pitched, snore-like croak. Often calls while submerged, which muffles the call. Male is smaller than the female and has enlarged thumbs on forefeet. Eggs are laid underwater and attach to aquatic vegetation. Moves to moist fields and meadows near water for the summer. Uses toxic secretions as a defense against predators. These also work against other amphibians, even if they just contact its skin.

Mink Frog
Rana septentrionalis

Frogs and Toads
(AMPHIBIAN)

Family: True Frogs (Ranidae)

Size: 2-3" (5-7.5 cm)

Description: Green to brown with dark blotches or mottling over most of body. Belly is white to pale yellow (venter), sometimes mottled with gray. Bright green upper lip. Few individuals have prominent dorsolateral folds. Webbing on hind feet extends to end of fifth toes.

Eggs/Young: 500-4,000 eggs in round masses or clusters once per year; tadpole; hatches into tadpole in 10-20 days, transforms the following year; some take as long as 2 years to transform; the warmer the water, the less time needed to hatch and transform

Origin/Age: native; 5-10 years

Habitat: rivers, streams, deep cold lakes

Overwinter: underwater; burrows into mud in lakes and ponds

Food: aquatic insects, snails, minnows, beetles

Compare: The Green Frog (pg. 127) has dorsolateral folds and webbing doesn't extend to the end of the fifth toes.

Stan's Notes: The most aquatic frog in Michigan. Produces a musky onion odor when handled similar to the scent of minks, hence the common name. The odor presumably helps repel predators. Emerges later than most frogs. While floating on the water's surface, male calls a low-pitched "tok-tok-tok" that sounds like someone is hitting two sticks together. Male is smaller than female, with larger tympanums (eardrums) than its eyes. Female tympanums are smaller than its eyes. Eggs are laid in water and often attach to plants. Tadpoles need clear, cold water for proper development. True Frogs family member, with more than 700 species worldwide except for Antarctica and Australia. Declining population due to changing climates (warming).

STATUS: Stable **123**

green morph

green brown mix

brown morph

Northern Leopard Frog

Rana pipiens

Family: True Frogs (Ranidae)

Size: 2-3½" (5-9 cm)

Description: Green or brown with 2-4 rows of dark round spots of irregular sizes on the back and sides. White belly. White chin. Pronounced dorsolateral folds. Light line on upper lip. Dark bars across hind legs. May have a slight green wash to inside thighs of hind legs.

Eggs/Young: 500-5,000 eggs in masses once per year; tadpole; hatches into tadpole in 7-28 days, most transform in 60-90 days; the warmer the water, the less time needed to hatch and transform

Origin/Age: native; 8-10 years

Habitat: wetlands, ponds, lakes, meadows, fields near water

Overwinter: underwater; burrows in mud at the bottom of lakes and larger ponds

Food: beetles, crickets and other insects, frogs, slugs, snails

Compare: Pickerel Frog (pg. 121) has 2 rows of squarish spots on the back and a yellow wash to inside thighs.

Stan's Notes: One of our most widespread and common frogs. Can give a loud scream or low vibrating buzz if seized. Emerges in April. Will travel up to a mile to get to breeding ponds. Frequently seen crossing roads with long leaping bounds. Breeding starts when water temperatures rise above 65°F (18°C). The male gives a low, guttural snore call lasting 2-4 seconds followed by several clucks and croaks. Moves from ponds to fields and meadows during summer if there is enough rain. Moves back to wintering lakes in fall and seen crossing roads. Populations decreased over the previous 30-50 years due to environmental toxins, habitat loss and overharvesting for biological supply and fishing bait.

STATUS: Stable **125**

male

female

Green Frog
Rana clamitans

Frogs and Toads
(AMPHIBIAN)

Family: True Frogs (Ranidae)

Size: 2-4" (5-10 cm)

Description: Bright green to brown, often with dark blotches on most of body. White belly, sometimes mottled with gray. Bright green upper lip. Prominent dorsolateral folds. Tympanums (eardrums) are larger than eyes in males and nearly equal to eyes in females. Hind feet are partially webbed. Male has a yellow vocal sac.

Eggs/Young: 1,000-5,000 eggs in floating masses once per year; tadpole; hatches into tadpole in 3-5 days, transforms in 90-120 days, in late summer or the following year; the warmer the water, the less time needed to hatch and transform

Origin/Age: native; 5-10 years

Habitat: permanent wetlands, rivers, streams, ponds, lakes

Overwinter: underwater; burrows into mud in ponds and lakes

Food: grasshoppers, crickets and other insects, slugs, snails

Compare: Mink Frog (pg. 123) produces a musky onion odor, usually lacks well-defined dorsolateral folds and has webbing to the end of the fifth toes on the hind feet.

tan's Notes: One of six species of True Frogs in Michigan. Lives in eep ponds and larger lakes that don't completely freeze. Lacks toxic kin secretions. Eaten by various animals. Often close to water's edge. umps in water and hides on bottom to escape predators. Emerges in arly May. Sings a single "plunk" like the pluck of a loose or out-of-une banjo string. Male is larger, with larger tympanums and thumbs. ggs laid in water attach to plants. Moves over land (mostly juveniles) o smaller ponds for the summer. Frogs overwintering in creeks that lon't freeze may be active on warmer days in late winter.

STATUS: Stable **127**

male

female

dark morph

Bullfrog
Rana catesbeiana

Family: True Frogs (Ranidae)

Size: 4-8" (10-20 cm)

Description: Varying shades of green from bright green to olive to nearly black. White-to-gray belly (venter), sometimes mottled with gray. Green upper lip. A fold of skin wraps around tympanums (eardrums) and extends to shoulders. Lacks dorsolateral folds. Hind feet are webbed except for last joint of largest toe. Male has a yellow vocal sac. Female often has a white throat.

Eggs/Young: 5,000-20,000 eggs in thin, floating film 3 feet (.9 m) wide once a year; tadpole; hatches into tadpole in 3-6 days, transforms in 2-3 summers; the warmer the water, the less time needed to hatch and transform

Origin/Age: native; 8-15 years

Habitat: permanent ponds, small lakes, river breakwaters

Overwinter: underwater; burrows into mud in ponds and lakes

Food: large insects, other frogs, crayfish, fish, small turtles, snakes, birds, small mammals

Compare: Adults are larger than most frog species. Green Frog (pg. 127) has dorsolateral folds down its back.

tan's Notes: Our largest frog. Rarely away from water. Home range of 0-40 feet (9-12 m). Loud scream when grabbed. Squeals, then hops r runs over water when scared. Male is larger than female, with larger ympanums than its eyes. Female tympanums are the same size as its yes. A warm weather frog, inactive until water is above 60°F (16°C). Emerges in late April. Male sits in water or on aquatic plant, singing loud, deep "rrum." Can be heard up to a quarter mile away. Bullfrogs way from water are often juveniles searching for their own ponds.

Fowler's Toad
Bufo fowleri

Frogs and Toads
(AMPHIBIAN)

Family: True Toads (Bufonidae)

Size: 2-3.5" (5-9 cm)

Description: A small, light tan-to-gray toad. Many irregular spots and blotches with 1 or many bumps. Often has a light stripe down the middle of back. Chin, throat and belly are white to dull yellow, sometimes with a few dark spots. Male has thickened dark thumbs, a gray throat (vocal) sac and is smaller than female.

Eggs/Young: 7,000-10,000 eggs in long gelatinous strings once per year; tadpole; hatches into tadpole in 2-7 days, transforms in 30-40 days; the warmer the water, the less time needed to hatch and transform

Origin/Age: native; 3-5 years

Habitat: sand prairies, wet meadows, beaches, suburban yards, ditches in farmlands, sandy river valleys

Overwinter: underground; burrows into earth below the frost line

Food: grasshoppers, beetles, other insects, spiders, worms

Compare: Very similar to American Toad (pg. 133), which is often larger, darker and has fewer bumps on its dark patches. Male Fowler's low nasal bleat is 2-7 seconds, unlike American's trill of 10-30 seconds or more.

Stan's Notes: A toad of sandy soils that is locally common in the right habitat. It is found in the Lower Peninsula along the eastern shore of Lake Michigan, while American Toad is found throughout the state. The range and moderating effects of the lake on local weather suggest it is less cold tolerant than the American. More active during daytime than the American. Recently considered a separate species. Besides puffing up, urinating and secreting a foul fluid from the parotoid glands, also feigns death as a defense. Won't cause warts if handled.

STATUS: Stable **131**

brown morph

American Toad
Bufo americanus

Frogs and Toads
(AMPHIBIAN)

Family: True Toads (Bufonidae)

Size: 2-4" (5-10 cm)

Description: Short hind legs, stocky body and bumpy skin. Can range from dark brown to tan (skin rarely red to olive green or light skin between dark patches). May have a light line along the middle of back (dorsum). Lighter brown belly (venter) with black spots. Brown-to-black splotches and spots, usually associated with bumps ("warts"). Two distinct ridges (cranial crests) between eyes. Large oval warts (parotoid glands) behind eyes.

Eggs/Young: 2,000-20,000 eggs in long gelatinous strings once per year; tadpole; hatches into tadpole in 4-10 days, transforms in 30-60 days; the warmer the water, the less time needed to hatch and transform

Origin/Age: native; 5-8 years

Habitat: woodlands, prairies, wetlands, grassy lawns, fields

Overwinter: underground; burrows to below the frost line

Food: grasshoppers, other insects, worms, spiders, slugs

Compare: Fowler's Toad (pg. 131) is often smaller, lighter and has more bumps on its dark patches. Male Fowler's low nasal bleat lasts 2-7 seconds, while American's higher musical trill lasts 10-30 seconds or more.

Stan's Notes: Our most common toad. A friend to gardeners, eating many insects. Changes from light to dark and back with temperature. Most active on warm, rainy evenings or nights. Uses hind legs to back into soft soil during days. A short hop or walk. If seized, will urinate, inflate itself with air and secrete a foul fluid from the parotoid glands. Will not cause warts if held. Male is smaller than female, with a gray throat patch and enlarged thumbs on forefeet during mating season.

GLOSSARY

Some of these glossary terms are depicted on the photographs on pages xii-xiv.

Advertisement call: A sustained call given in a relatively organized manner by a male frog or toad (rarely by the female of some species) to attract a mate for breeding. Usually is referred to as a call. Sometimes known as a song.

Anal plate: The divided or single scale or plate just before and covering the anal opening of a snake. See *cloaca* and *vent*.

Aquatic: Living in water or pertaining to the water as opposed to *terrestrial*.

Belly scales: The large, wide scales on the underside of a snake from just below the head to the vent, typically extending across the belly from side to side.

Carapace: The upper shell as opposed to the plastron or lower shell of a turtle.

Caruncle: A tiny, pointed, tooth-like structure on the upper mandible of hatchling turtles, used for breaking open the eggshell from within, similar to a hatching bird.

Cloaca: The opening in an animal where urinary, intestinal and reproductive substances exit the body. The anus, often called the vent. See *anal plate*.

Costal groove: Vertical grooves on the sides of a salamander, each corresponding to the space between the ribs.

Cranial crests: The bony bumps or ridges between the eyes of some toad species. Not the same as the *parotoid glands*.

Crepuscular: Active during the early morning and late evening hours as opposed to day or night. See *diurnal* and *nocturnal*.

Diurnal: Active during daylight hours as opposed to nighttime hours. Opposite of *nocturnal*.

Divided anal plate: An anal plate of a snake that is divided into two parts.

Dorsum: The back of an animal. Opposite of *venter*.

Dorsolateral fold: An elevated ridge of skin that runs lengthwise down the sides of the back of some frog species.

Ectotherm: An animal that obtains its body heat from the surrounding environment rather than regulating its body temperature internally.

Eft: The terrestrial sub-adult form of a newt, often orange or red in color.

Estivation: A time of inactivity that is usually associated with a seasonal drought or excessive heat such as occurs in summer. Opposite of *overwinter*.

Forefeet: The front feet as opposed to the hind feet or rear feet of an animal.

Gravid: A reproductive condition of carrying eggs or being pregnant with live young.

Hatchling: The young of a reptile or amphibian that has recently hatched out of its egg.

Heat pit: The depression on each side of the face of venomous snakes in between the eye and nostril, where heat-sensitive organs are located. Also called facial pit.

Herpetology: The scientific study of reptiles and amphibians.

Hind feet: The rear feet as opposed to the forefeet or front feet of an animal.

Juvenile: The stage between hatching or transforming and a breeding adult. Some reptile and amphibian juveniles differ from the adults of their species in color and size.

Keel: A prominent ridge on the carapace or upper shell of a turtle, usually running lengthwise down the center or middle (median). Often saw-toothed.

Keeled scales: Scales of a snake that have a ridge or keel along the length of each scale, making them feel rough to touch, as in the Common Garter Snake. See *smooth scales*.

Larva: The usually aquatic, immature amphibian form, often with a large round head and long fin-like tail, that uses gills to breathe in water. See *tadpole*.

Lateral: On the sides or pertaining to the sides of an animal.

Marginal scutes: The row of scutes that make up the outer edge of a hard-shelled turtle's carapace or plastron. Marginal scutes are smooth, wavy, toothed or a combination.

Median: In the middle or pertaining to the center of the back of an animal.

Melanistic: A condition of increased dark pigment (melanin), resulting in darkened shells, scales, skin or eyes.

Morph: The same species occurring in several well-established and permanent colors, shapes or structural differences.

Nocturnal: Active during nighttime hours as opposed to daylight hours. Opposite of *diurnal*.

Overwinter: The act of a reptile or amphibian surviving winter. The equivalent to hibernation in some mammals. Opposite of *estivation*.

Oviparous: Egg laying or egg depositing.

Ovoviviparous: Non-egg laying. Eggs are hatched within the body, which results in the birthing of live young.

Parotoid glands: The wart-like glands (swollen bumps) on top of the head of toads just behind the eyes that produce a toxic substance used for defense or to repel predators.

Plastron: The lower shell as opposed to the carapace or upper shell of a turtle.

Population: All individuals of a species in a specific area.

Predator: An animal that hunts, kills or eats other animals.

Prey: An animal that is hunted, killed or eaten by a predator.

Scale rows: The horizontal rows of scales on the sides (lateral) of a snake. A numbering system identifying scale rows starts the first scale row immediately above the belly scales and goes up toward the middle of the back (dorsum).

Scutes: The hard overlapping scales that form the upper and lower shells of a hard-shelled turtle. Sometimes called plates.

Single anal plate: Anal plate of a snake that is not divided into two parts. Also called undivided anal plate.

Smooth scales: Scales of a snake that lack a ridge or keel along the length of each scale, making them feel smooth to touch, as in the Smooth Green Snake. See *keeled scales*.

Snout: The nose or pertaining to the tip of the nose of an animal.

Tadpole: The aquatic larval form of frogs and toads. See *larva*.

Terrestrial: Living on land or pertaining to the land as opposed to *aquatic*.

Toe pads: Enlarged, sticky tips of fingers and toes in treefrogs that adhere on contact to a variety of surfaces.

Tympanums: The visible disk-like membranes usually behind the eyes in frogs and turtles. The eardrums.

Vent: The external opening of the cloaca. See *anal plate*.

Venter: The belly of an animal. Opposite of *dorsum*.

Vocal sac: A region on or near the throat of male frogs and toads that inflates with air, amplifying sound (resonating chamber).

HELPFUL RESOURCES

Snakebite Emergency
For a snakebite, please seek medical attention at an emergency room, call 911 or contact the following poison control center, which serves Michigan statewide:

Regional Poison Center
800-764-7661

Web Pages
The Internet is a valuable place to learn more about reptiles and amphibians. You may find studying herps on the Net a fun way to discover additional information about them or to spend a long winter night. These sites will assist your pursuit of herpetology.

Site and Address
Michigan Society of Herpetologists
www.michherp.org/main1.html

Michigan DNR
www.michigan.gov/dnr

Author Stan Tekiela's homepage
www.naturesmart.com

AUDIO CD TRACK INDEX

Part III

Conclusion

CHECK LIST/INDEX

Use the boxes to check the reptiles and amphibians you've seen.

ABOUT THE AUTHOR

Stan Tekiela is a naturalist, author and wildlife photographer with a Bachelor of Science degree in Natural History from the University of Minnesota. He has been a professional naturalist for more than 20 years and is a member of the Minnesota Naturalist Association, Minnesota Herpetological Society, Outdoor Writers Association of America and Canon Professional Services. Stan actively studies, photographs and records frogs, toads and birds throughout the United States. He received an Excellence in Interpretation award from the National Association for Interpretation and a regional award for Commitment to Outdoor Education. A columnist and radio personality, his syndicated column appears in over 20 cities and he can be heard on a number of radio stations. Stan resides in Victoria, Minnesota, with wife Katherine and daughter Abigail. He can be contacted via his web page at www.naturesmart.com.

Stan authors other field guides for Michigan including guides for birds, wildflowers and trees.